Indian Ants

The Authors

Dr. Sathe Tukaram Vithalrao [M.Sc., Ph.D., Sangit Vishard, IBT (Seri.), F.I.S.E.C., F.S.E.Sc., F.S.L.Sc., F.I.C.C.B., F.S.S.I., FHAs] is presently working as Professor and Former Head, Department of Zoology, Shivaji University, Kolhapur. He has teaching experience of 32 years in Entomology at University PG department and 21 years in Agrochemicals and Pest Management. He has written 40 books and published 355 research papers in national and international journals of repute. He guided 29 Ph.D. students and completed 9 major research projects (from CSIR, DST, DBT and UGC). He visited Canada (1988), Japan (1988), Thailand (2002, 2004), Spain (2005), France (2005), South Korea (2006) and Nepal (2007) etc. for academic work. He is member of editorial board of 13 prestigious journals. He delivered 35 talks through All India Radio and international conferences and involved in Doordarshan, S.T.V. and B.T.V. programmes on useful and harmful insects. He published more than 35 popular articles in daily newspapers on insects and sericulture. He got several prestigious awards like "Environmentalists of the Year-2003", "Bharat Jyoti", "Jewel of India", "International Gold Star", "Eminent Citizen of India", "Education Acumen", "Best Educationist", "Eminent Scientist of the Year-2008", "Lifetime Education Achievement", "Lifetime Achievement in Zoology (Insect Taxonomy)-2009", Education Leadership-2011, Asia Pacific International Award-2012, Global Education Leadership Award-2013, Indo-Nepal Shiromani Award, etc. He is also working as Research and Recognition (RR) Committee member for Pune University, Pune; North Maharashtra University, Jalgaon; Shivaji University, Kolhapur and DBA Marathwada University, Aurangabad. He has been awarded several fellowships from different scientific and academic societies. He is Chairman of Maharashtra District Environmental Centre of NESA.

Dr. (Smt.) Shilpa Haribhau Kurane (M. Sc; Ph. D.) is Research Scientist in Department of Zoology, Shivaji Univarsity, Kolhapur. She has published 10 research papers in various journals of international repute with high impact factor. She is contributory teacher for graduate course in zoology and attended and presented several research papers in national conferences.

Dr. P.M. Bhoje (M.Sc., Ph.D.), Head, Department of Zoology, Y.C. Warna College, Warnanagar, Kolhapur, Maharashtra has published 25 research papers in various journals of national and international repute. He is BOS member (zoology) of Shivaji University, Kolhapur and guides 2 Ph.D. students. He visited Thailand and Sri Lanka for presentation of research papers in conferences and President of 'Green Guards' Society, Kolhapur, and NESA district Centre, Kolhapur. He has written 2 books and shared popular chapters in several edited books.

Indian Ants

– *Authors* –

Prof. (Dr.) T.V. Sathe

Dr. (Smt.) Shilpa Haribhau Kurane

Dr. P.M. Bhoje

2017

Daya Publishing House®

A Division of

Astral International Pvt. Ltd.

New Delhi – 110 002

Cataloging in Publication Data--DK
Courtesy: D.K. Agencies (P) Ltd. <docinfo@dkagencies.com>

Sathe, T. V., author.
Indian ants / authors, Prof. (Dr.) T.V. Sathe, Dr. (Smt.) Shilpa Haribhau Kurane, Dr. P.M. Bhoje.
pages cm
Includes bibliographical references and index.
ISBN 9789389605136 (Int Edition)

1. Ants--India--Maharashtra--Classi ication. I. Kurane, Shilpa Haribhau, author. II. Bhoje, P. M. (Prakash Maruti), 1965- author. III. Title.

QL568.F7S28 2017 DDC 595.796095479 23

Published by : **Daya Publishing House®**
A Division of
Astral International Pvt. Ltd.
– ISO 9001:2008 Certified Company –
4760-61/23, Ansari Road, Darya Ganj
New Delhi-110 002
Ph. 011-43549197, 23278134
E-mail: info@astralint.com
Website: www.astralint.com

Digitally Printed at : **Replika Press Pvt. Ltd.**

Preface

Ants belong to Order Hymenoptera, containing about 9000 described species from the world. They are social insects and found in all terrestrial habitats. Ants are visualized as pests of agricultural, horticultural and forest plants and used in biological pest control of insect pests. Ants have nutritional and medical importance. Hence knowledge on their diversity, seasonal abundance, distribution and host records will add great relevance for understanding ants for their management and their utility in biological pest control.

The present work contain 24 described species of ants with their seasonal abundance and distribution in Western Maharashtra (Kolhapur, Sangli, Satara). I feel that the book will helpful for students, teachers and scientists in the field of Entomology and Environmental sciences.

Prof. (Dr.) T.V. Sathe

Dr. (Smt.) Shilpa Haribhau Kurane

Dr. P.M. Bhoje

Contents

1
Introduction

India is bounded by Indian Ocean in south-west the Arabian Sea and in south-east Bay of Bengal and is very good place for diversity of wildlife in different protected habitats. According to Dreeze and Sen (1995) India made significant progress in human development and modernization. India, a mega diverse country with only 2.4 per cent of the world's land area, accounts for 7-8 per cent of all recorded species, including over 45,000 species of plants and 91,000 species of animals (MoEF, 2014). Being one of the 17 identified mega diverse countries, India has 10 biogeographic zones and is home to 8.58 per cent of the mammalian species documented so far, with the corresponding figures for avian species being 13.66 per cent, for reptiles 7.91 per cent, for amphibians 4.66 per cent, for fishes 11.72 per cent, for plants 11.80 per cent (MoEF,2014) and for insects 75-80 per cent (Alfred *et al.*, 1998). Four of the 34 globally identified biodiversity hotspots, namely the Himalaya, Indo-Burma, the Western Ghats-Sri Lanka and Sundaland are represented in India (MoEF, 2014). India is also potential centre of crop diversity and harbor hundreds of various crop plants such as rice, maize, millets etc. The diverse physical features and climatic conditions of the country have resulted in a variety of ecosystems such as forests, grasslands, wetlands, desert, coastal and marine ecosystems which harbour and sustain high biodiversity and contribute to human well-being (MoEF, 2014).

India is (Figure 1) amongst the few countries that have developed a biogeographic classification based on which conservation planning has been taken up (Rodgers *et. al.*, 2002). This classification uses four levels of planning units: the biogeographic zone, the biotic province, the land region and the biome (www.wii.gov.in). Within India, the biogeographic classifications recognize 10 zones divided into 27 provinces (Roy *et al.*, 2012). According to MoEF (2014) North Eastern India, the Andaman and Nicobar Islands and the Western Ghats as well as some patches of the Eastern Ghats (especially Araku Valley, Andhra Pradesh) were classified as having maximum biological richness.

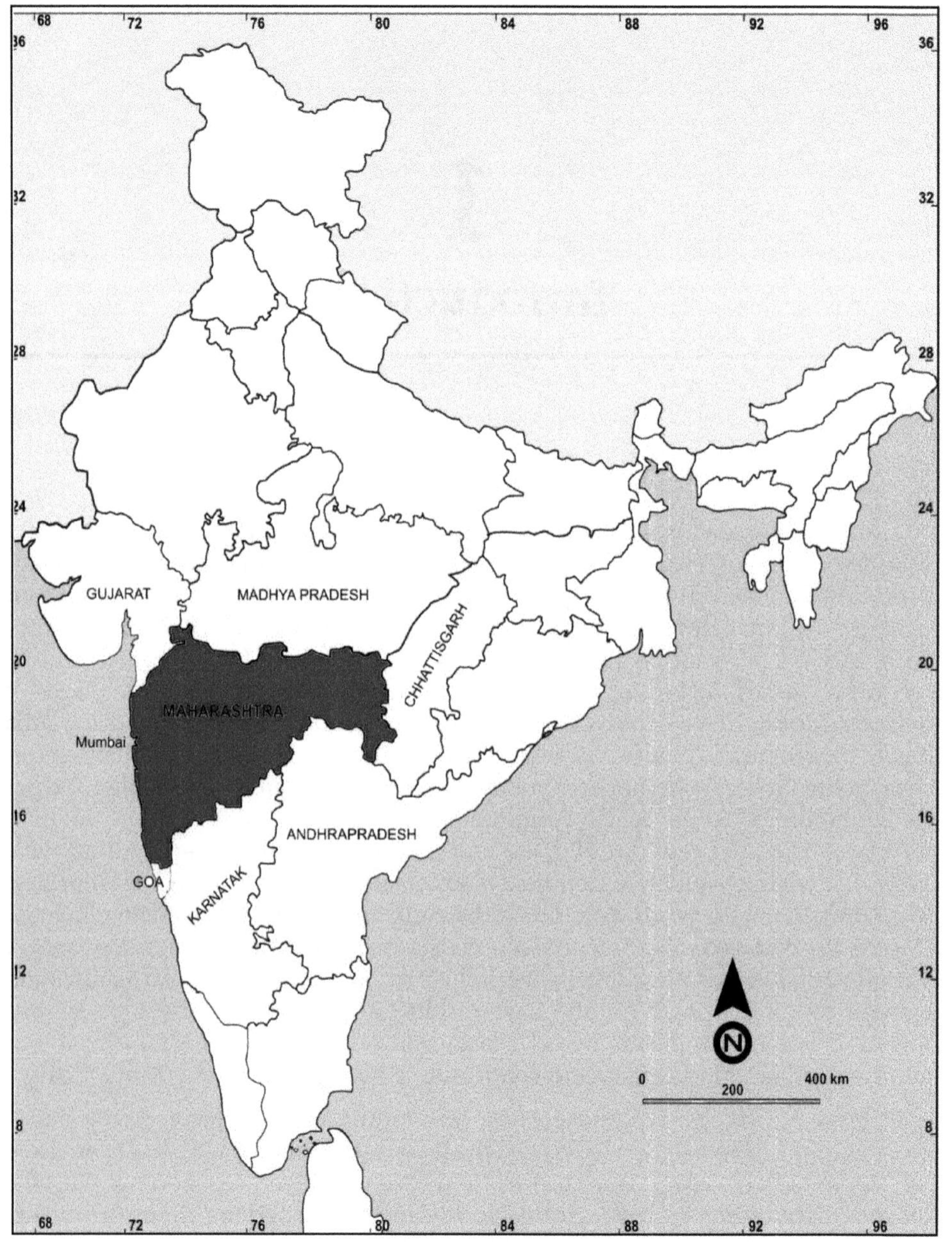

Figure 1: Map of India Showing Maharashtra.

The main natural habitat types for insects and other animals are: Forests, Grasslands, Wetlands, Mangroves, Coral reefs, Deserts (UNEP,2001).According to the forest survey of India assessment (1997) the forest cover of India is placed at 633 397 sq. km with 19.27 per cent of India's total geographical areas. India is endowed with diverse forest types ranging from the Tropical wet evergreen forests in north eastern to the Tropical thorn forests in the Central and Western India. The forests of India can be divided into 16 major groups comprising 221 types (Olson *et al.,* 1983).

The idea of hotspots was first introduced in 1988 by ecologist Norman Myers, who defined a hotspot as an area of exceptional plant, animal and microbe wealth that is under threat (MoEF, 2014). The key criteria for determining a hotspot is endemism (the presence of species found nowhere else on earth) and degree of threat. The Western Ghats (Figure 2) region is most important biogeographic zone of India which is richest centre of endemism.

Western Ghats extends from southern tip of Peninsula (8°N) to mouth of Taphi River (21° N) at altitudes between 900 to 1500 m above sea level. 'Palghat gap' which is almost the central part of Western Ghats located in Kerala state having very low altitude, less than 144 m above sea level (Joseph, 2004). The Western Ghats is scattered in five states of India.

The area of Western Ghats in Maharashtra is scattered in nine districts namely, Dhule, Thane, Raigad, Pune, Satara, Sangli, Kolhapur, Sindhudurg and Ratnagiri of which Thane, Raigad, Ratnagiri and Sindhudurg are the parts of Western slope while Dhule, Pune, Satara, Sangli and Kolhapur are the parts of eastern slope of the Western Ghats. The Western Ghats region also shows stepped appearance as well as steep escarpment zones. The former characterizes the flat surfaces to west of the Sahyadries of the Nashik district (at 300-360 m) while latter features are valley heads of the major streams, river Krishna at Mahabaleshwar, Indrayani and Ulhas at Lonavala, Bhima at Bhimashankar, *etc.* These varied geomorphic features are characterized by their typical altitude, climate, rainfall and rock types. These regions, therefore, are known for their characteristic forest zones such as Southern Tropical Semi-evergreen forests of Bhimashankar, Radhanagari and Amboli; Southern Tropical Moist Deciduous forests of Pein and Surgana Talukas of Nashik district, Wada and Jawahar in Thane district and Melghat region. Thorny forests are scattered from Khandesh in the north to the Solapur and Sangli districts in the north (BVIEER, 2010).

The annual flow pattern of river in the Western Ghats is in accordance with the monsoon rainfall. Most rivers are characterized by a tripartite sequence of flows (Diddee, 2002). Climatic conditions in the Western Ghats vary with the altitude and physical proximity to the Arabian Sea and the equator. Although the Western Ghats experience a tropical climate - being warm and humid during most of the year with mean the temperature ranging from 20°C in the south to 24°C in the north, the higher elevations experience subtropical climates and on occasions frost. The coldest periods in the Western Ghats coincide with the wettest (Daniels, 2001). Rainfall peaks at 9000 mm and above per year and as low as 1000 mm are frequent in the east bringing the average to around 2500 mm. The northern Western Ghats

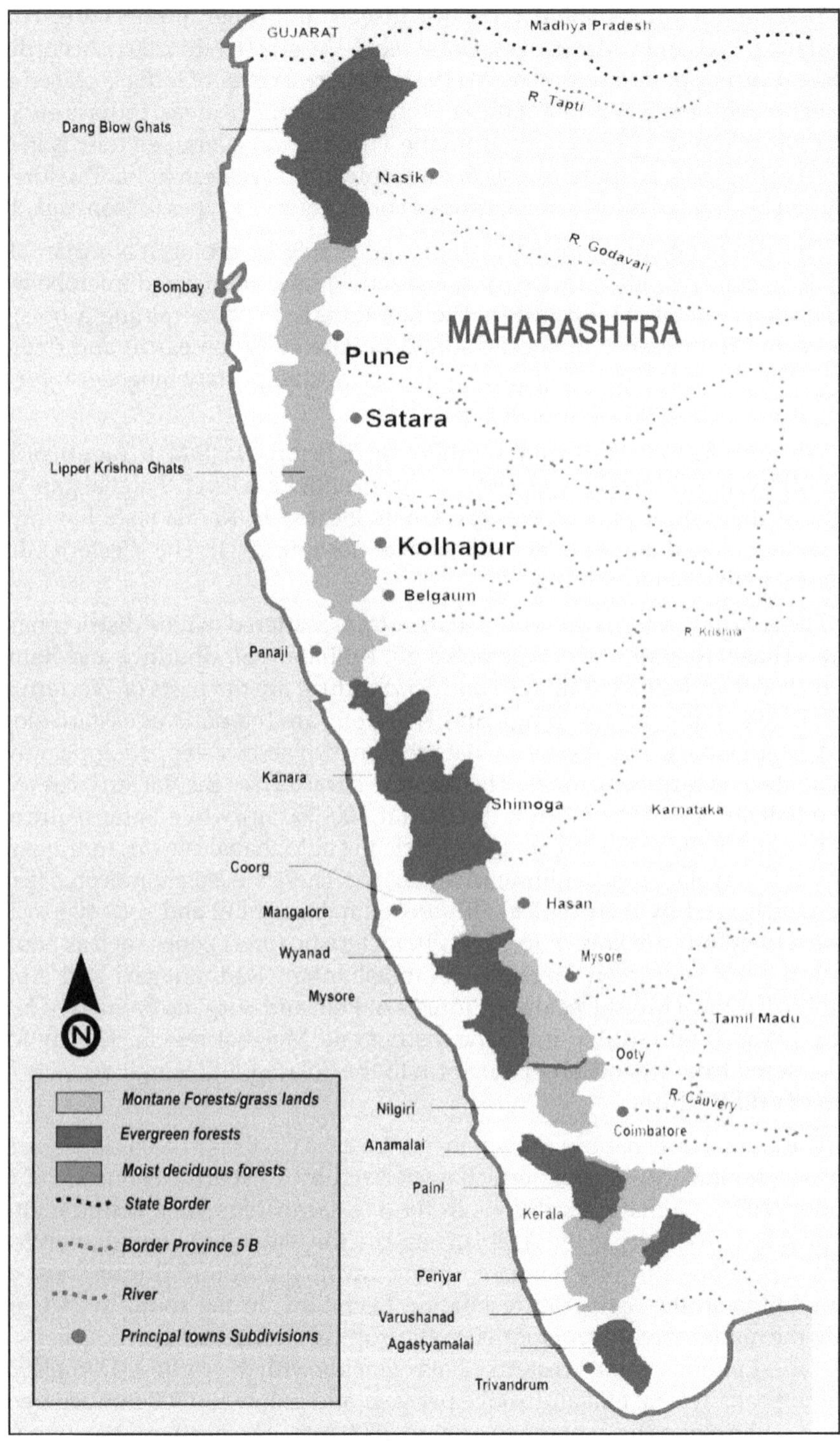

Figure 2: Biographic Subdivision of the Western Ghats.

receive the highest rainfall (locally over 9000 mm) and also shows dry weather over more than half the year (Daniels, 2001).

The Western Ghats is characterized by a very rich variety of fauna but, many of them are endemic (Rodgers and Panwar, 1988; Pandharbale and Sathe, 2008). Among the invertebrate groups, about 330 butterfly species (11 per cent endemic), 350 ant species (20 per cent endemic) and 174 odonate species (dragonflies and damselflies (40 per cent endemic) have been reported from the Western Ghats(Rodgers and Panwar 1988). Rodgers and Panwar (1988) listed many important species and groups of species found in Ghats. Fauna of Gujarat and Goa are published in 2000 and 2008 by Zoological Survey of India (ZSI). Similarly, a catalogue of new taxa described by ZSI during 1916-1991 is known to the science (Das, 2003).

According to Gaston and Spicer (2004) biodiversity is a variety of life includes variations of biological organization from genes to species to ecosystems which includes a number of components.

In Indian forests, large number of insect species has been recorded, 67,000 species of insects and many more are described by the scientist of ZSI (1983) and several others. Beason (1941) worked on large number of forest insects. He studied 3378 insects belonging to Orders Collembola, Thysanura, Ephemeroptera, Odonata, Orthoptera, Anaplura, Dermaptera, Isoptera, Coleoptera, Hemiptera, Thysanoptera, Hymenoptera, Lepidoptera, Neuroptera and Diptera. Chatterjee and Misra (1975) reported 680 species of parasitoids as natural enemies of insects.

Several prominent butterflies are found in Western Ghats. A total of 314 species of butterflies have been recorded from Western Ghats (Nair *et al.*, 1973). In 1979-80, ZSI studied the insects of silent valleys and reported 242 species. Joseph (1984) studied the insect life and ecodevelopments of Western Ghats. Sathe *et al.* (1986; 1987) studied the fauna of butterflies from Western Ghats of Maharashtra. They reported 33 species out of which 8 species are rare. Gaonkar (1996) listed 330 species of butterflies from Ghats region. Satish (1996) reported 13 species of moths and 16 species of butterflies from Shimoga. Sathe (1987) studied the fauna of butterflies from Western Ghats. Sathe (1992) studied the fauna of aphids on plants of economic importance found in Western Maharashtra. Kunte (1997) worked on seasonal patterns and abundance of butterflies in North and Western Ghats. Sathe and Shinde (2006) also studied the diversity of butterflies from Western Ghats.

Inamdar (1990) studied braconid parasitoids of some economic important crop pests from Western Maharashtra and described 12 new species and 2 species have been redescribed. Dawale (1991) studied natural enemies of economic important crop pests with special reference to Hymenoptera and described 9 species of braconid, 6 new species from Ichneumonids genera and 1 species have been redescribed. Ingawale (1991) studied microgasterinae of some important economic crop pests and described 8 new species. Bhosale (1994) studied Indian Coccinellidae with special reference to predacious species and described 17 new species, 2 species have been redescribed. Sathe and Mulla (1995) reported 17 insect pest species feeding on mulberry plants in Amboli of Western Ghats. Rokade (1996) studied diversity and distributional record of Braconids from Maharashtra and described 9 new species, 1 species have been redescribed.

Sathe *et al.* (1997) reported the diversity of silkworms in Western Maharashtra. Sathe and Pandharbale (1999) described 13 species of Hawk moths belonging to 6 genera from Western Ghats of Maharashtra. Pandharbale and Sathe (2001) reported a new species of genus *Syntomis* from Western Ghats. Patil (2002) studied insect predators of agricultural pests of Kolhapur district and described 20 new species, 1 species have been redescribed. Sathe and Pandharbale (2003) described biodiversity of moths from Western Ghats of Satara district of Maharashtra. Kadam (2014) studied biosystematics of Lepidoptera from Western Maharashtra and described 13 new species while 10 species have been redescribed. Sathe (2007) described biodiversity of wild silk moths from Western Maharashtra. Recently Sathe (2014) studied harmful syntomids of agro and forest crop plants from Western Maharashtra.

Kavane *et al.* (2003) studied natural enemies of tasar silk worm, *Antheraea mylitta* D. from Satara district of Maharashtra. Jadhav and Sathe (2006) described biodiversity of aphids from Satara district of Western Ghats. Jadhav and Sathe (2006) described biodiversity of aphids from Poona district of Western Ghats.Sathe and Shinde (2007) reported a new species of the genus *Crocothemis* Brauer from Western Ghats of Maharashtra. Sathe and Jadhav (2007) described a new species of genus *Aphis* from Western Ghats of Maharashtra. Gaikwad *et al.* (2007) studied seasonal distribution pattern of borrowing dung beetles. Sathe and Chougale (2008) investigated the biodiversity of termites from Western Ghats of Sindhudurg district. Sathe and Awate (2008) studied taxonomy of crickets from Kolhapur district and described 4 new species of genus *Gryllotalpa* and 5 new species of genus *Gryllus* have been redescribed. Sathe and Shinde (2008) studied the biodiversity of dragonflies from orus of Sindhudurg district. Diversity of Ichneumonid flies from Amba Ghats have been reported by Bhoje and Sathe (2008). Jadhav and Sathe (2008) described a new species of the genus *Aphis* (Aphididae) from Western Maharashtra. Chougale (2009) studied biosystematics of Ichneumonid and Braconid parasitoids from agroecosystems and described 3 new species of Braconids and 3 new species of Icheumonids. Sathe and Jagtap (2009) exposed tree hole breeding and resting of mosquitoes from Western Ghats of Maharashtra. Gaikwad *et al.* (2009) reported diversity of butterflies in Amba reserved forest. Sathe and Bhusnar (2010) studied biodiversity of dragonflies from Sawantwadi region of Western Ghats. Similarly, Sathe (2010) studied biodiversity of Damselflies from Koyana dam and around area. Sathe and Jagtap (2010) also studied biodiversity of Anopheline mosquitoes of Western Ghats.

Sathe and Chougule (2010) reported a synoptic list of Gerromorpha from Western Ghats of Sindhudurg district. Sathe *et al.* (2010) highlighted dragonflies of Western Ghats of Sindhudurg district. Sathe (2011) described ecology of mosquitoes from Kolhapur district. Jagtap (2011) studied systematic and molecular phylogeny of mosquitoes and described 13 new species, 3 species have been redescribed. Bhusnar and Sathe (2011) described a biodiversity of grasshoppers from Kolhapur district. Bhusnar and Sathe (2012) explained the ecology of grasshoppers from Kolhapur district. Bhusnar (2012) studied diversity of grasshoppers and described

4 new species, 50 species have been redescribed. Patil (2012) studied chromosomal biodiversity in Coleopterous beetles (Coccinellidae) and recorded 19 species of lady bird beetles out of which 9 were new species. Sathe (2012) studied biodiversity of Tachinid flies from Western Maharashtra. Sathe (2012) also studied biodiversity of Ichneumonids flies from Western Ghats. Sathe *et al.* (2013) reported diversity of dipterous forensic insects from Western Maharashtra. Sathe and Patil (2014) studied diversity and utility of toads in biological insect pest control in Satara. Bhoje *et al.* (2014a) studied diversity of ants from Kolhapur district. Bhoje *et al.* (2014b) studied biodiversity of ants of Amba reserve forest of Western Ghats. Jadhav *et al.* (2014) studied rearing performance of Tasar silk worm *Antheraea mylitta* Drurry on different food plants from Kolhapur district.

Sathe and Shinde (2014) studied biodiversity, abundance and predatory status of Odonates from paddy ecosystems of Kolhapur district. Desai and Sathe (2014) studied destructive Cecidomyids of forest trees of Western Ghats. Sathe (2014) reported ecology, epidemiology and control of sand flies from Kolhapur region. Sathe (2014) also studied biodiversity and biology of Cerambycid beetles from Western Maharashtra. Nine new species of mantids and their insect pest predatory potential from agro ecosystems of Kolhapur have been reported by Sathe and Patil (2014). Sathe and Chougale (2014) studied hymenopterous bio pesticides and their preliminary biocontrol potential from Western Maharashtra including Ghats. Jadhav Divya and Sathe (2014) studied altitudinal diversity of forensic Blowflies of Western Ghats. Sathe *et al.* (2014) studied floral host plants for Tachinid flies from Kolhapur and Satara district. Recently, Sathe and Gangate (2015) studied host plants for a whitefly *Aleurodicus dispersus* R. from Kolhapur region. Khairmode *et al.* (2015) studied biology, ecology and control of weevils on banana from Kolhapur region. Desai *et al.* (2015) studied ecology and ethology of crane fly *Tipula palndosa* Meign from Kolhapur region. Sathe (2015) recorded insects for human diet from Kolhapur region including Ghats while, Sathe *et al.* (2015) studied poisonous insects of Kolhapur including Western Ghats.

According to Gadgil *et al.* (MoEF Report,2011) among the invertebrate groups, about 350(20 per cent endemic) species of ants, 330(11 per cent endemic) species of butterflies, 174(40 per cent endemic) species of odonates (dragonflies and damselflies) and 269 (76 per cent endemic) species of molluscs (land snails) have been described from Western Ghats.

The perusal of literature indicates that many insect groups are attempted from Western Ghats. However, little attention is paid on ants.

Ants (Hymenoptera : Formicidae) are eusocial insects which are evolved from wasp-like ancestors in the mid-Cretaceous period between 110 and 130 million years ago and diversified after the rise of flowering plants. More than 12,500 of an estimated total of 22,000 species have been classified (Agosti and Johnson, 2003).

Ants are easily identified by their elbowed antennae and the distinctive node-like structure that forms their slender waists. Ants form colonies that range in size from a few dozen predatory individuals living in small natural cavities, highly organized colonies that may occupy large territories and consist of millions of individuals. According to Oster and Wilson (1978) larger colonies consist mostly of sterile, wingless females forming castes of "workers", "soldiers" or other specialized groups. Nearly all ant colonies also have some fertile males called "drones" and one or more fertile females called "queens".

Ants are found in almost every landmass on Earth. However, they are lacking in Antarctica and a few remote or inhospitable islands. According to Schultz (2000) ants thrive in most ecosystems and may form 15–25 per cent of the terrestrial animal biomass. Their success in so many environments has been attributed to their social organization and their ability to modify habitats, tap resources and dependence against enemies. Their long co-evolution with other species has led to mimetic, commensal, parasitic, and mutualistic relationships (Holldobler and Wilson,1990).

Ants have division of labour and an ability to solve complex problems between them (Dicke *et al.*, 2004). Many human cultures make use of ants in cuisine, medication and rituals. Some species act as biocontrol agents (Holldobler and Wilson, 1990). Ants are herbivores, predators, and scavengers. They occupy a wide range of ecological niches. However, most species are omnivorous generalists, but a few are specialist feeders (Schultz, 2000). Ants measures from 0.75 to 52 millimeters in body length (Shattuck, 1999), the largest species being the fossil *Titanomyrma giganteum* L. the queen of which is 6 centimeters (2.4 in) long with a wingspan of 15 centimeters (5.9 in) (Schaal, 2006).

There is colour variation in ants. Most ants are red or black, but a few species are green and some tropical species have a metallic luster. The life stages of ant consist of 4 distinct stages *viz.* egg, larva, pupa and adult. Fertilized eggs produce females (diploid); if not, it will be male (haploid). The larva is largely immobile and is fed and cared for by workers. Larvae, especially in the later stages, may also be provided solid food such as trophic eggs, pieces of prey, and seeds brought by workers. The larvae grow through a series of four or five moults and enter the pupal stage. The pupa is exarate type; the appendages are free and not fused to the body (Gillott, 1995; Holldobler and Wilson, 1990).

The queens can survive for 13 years, and workers for 1 to 3 years. Males, quite short-lived and surviving for only a few weeks (Keller, 1998). For communication ants can use pheromones, sounds, and touch (Jackson and Ratnies, 2006). The use of pheromones as chemical signals is more developed in ants. Ants perceive smells with their long, thin, and mobile antennae. According to Goss *et al.* (1989) the paired antennae provide information about the direction and intensity of scents.

Ants attack and defend themselves by biting and, in many species, by stinging, often injecting or spraying chemicals, such as formic acid in the case of Formicinae ants, alkaloids and piperidines in fire ants, and a variety of protein components in other ants. Bullet ants (*Paraponera*), located in Central and South America, are considered to have the most painful sting of any insect, although it is usually not fatal to humans. This sting is given the highest rating on the Schmidt Sting Pain Index. The sting of jack jumper ants can be fatal (Clarke, 1986), and an antivenom has been developed for it (Brown *et al.*, 2005). Fire ants, *Solenopsis* spp. are unique in having a poison sac containing piperidine alkaloids (Obin and Vander Meer, 1985). Their stings are painful and can be dangerous to hypersensitive people (Stafford, 1996).

Nesting in many ants is complex build however; other species are nomadic and do not build permanent nests. Ants may form subterranean nests or build them on trees. Ant nests may be found in the ground, under stones or logs, inside logs, hollow stems, or even acorn. The materials used for construction of nest include soil and plant matter (Holldobler and Wilson, 1990). According to Franks *et al.* (2005) ants carefully select their nest sites; *Temnothorax albipennis* C. will avoid sites with dead ants, for avoiding pests and diseases.

According to Wilson and Holldobler (2005) most ants are generalist predators, scavengers, and indirect herbivores but a few have evolved specialized ways of obtaining nutrition. Many ant species that engage in indirect herbivore rely on specialized symbiosis with their gut microbes (Anderson *et al.*,2012) to upgrade the nutritional value of the food they collect (Feldhaar *et al.*,2007) and allow them to survive in nitrogen poor regions, such as rainforest canopies (Russell *et al.*,2009). Workers are specialized in related tasks and their size. The largest ants cut stalks, smaller workers chew the leaves and the smallest tend the fungus. Symbiotic bacteria on the exterior surface of the ants produce antibiotics that kill bacteria introduced into the nest that may harm the fungi (Schultz, 1999). Ants form symbiotic associations with other ant species, other insects, plants, and fungi. Ants are predated by many animals and even certain fungi. Aphid-ant and mealy-bug ant associations are well known symbiotic relationships. Aphids and other hemipteran insects secrete a sweet liquid called honeydew, while feed on plant sap. The sugars in honeydew are a high-energy food source, which many ant species collect. In some cases, the aphids secrete the honeydew in response to ants tapping them with their antennae. The ants in turn keep predators away from the aphids and will move them from one feeding location to another. When migrating to a new area, many colonies will take the aphids with them, to ensure a continued supply of honeydew. Ants also tend mealy bugs to harvest their honeydew. Mealy bugs may become a serious pest of pineapples if ants are present to protect mealy bugs from their natural enemies (Styrsky and Eubanks, 2007; John and Beardsley, 1996).

Ants play an important role in dispersal of seeds of many tropical tree species (Hanzawa *et al.*, 1998; Giladi, 2006). Some plants in fire-prone grassland systems are particularly dependent on ants for their survival and dispersal as the seeds are transported to safety below the ground. Many ant-dispersed seeds have special external structures, elaiosomes that are sought after by ants as food (Fischer *et al.*, 2005).

Ants perform many ecological roles that are beneficial to humans, including the suppression of pest populations and aeration of the soil. (Holldobler and Wilson, 1990). On the other hand, ants may become nuisance when they invade buildings, or cause economic losses. In some parts of the world (mainly Africa and South America), large ants, especially army ants, are used as surgical sutures. The wound is pressed together and ants are applied along it (Sapolsky, 2001).

Some ants have toxic venom and are of medical importance. (Haddad *et al.*, 2005; McGain and Winkel, 2002). Many species of ants are edible. Their eggs, larvae and pupae are eaten in different parts of the world (DeFoliart, 1999; Sathe, 2015).

Some species of ants are also visualized as pests (Atwal, 1978, Sathe, 2013). The presence of ants can be undesirable in places meant to be sterile. They can also come in the way of humans by their habit of raiding stored food, damaging indoor structures, causing damage to agricultural crops either directly or by aiding sucking pests or because of their stings and bites (Rust and Choe, 2007).

In Maharashtra there are various agro and forest ecosystems in Kolhapur, Sangli and Satara districts.Western Ghats is one of the most important hotspot of the biodiversity. Keeping in view the versatile nature of ants and their different economic roles, the present topic was selected.

2

Review of Literature

Ants are eusocial hymenopterous insects and have great values as object of study. The abundance of ants and its ecological dominance are shown by their high degree of variability as exhibited in the great number of species, subspecies and varieties and by their extraordinary geographical ranges. The local diversity of ants is substantial and occupying a wide range of feeding niches in the vegetation and soil. Ants constitute about 15 per cent of the animal biomass (Fittkau and Klinge, 1973). They are called as herbivores because they harvest about 15 per cent of the herbivore in tropical forests. Erwin (1989) at Peru showed that 69 per cent insect specimens collected by fogging the forest canopy were ants.

Ants belong to the family Formicidae. Formicidae is one of the largest family of order Hymenoptera of class Insecta and widely distributed throughout the world as cosmopolitan individuals. About 10,000 species of ants under 16 sub-families, 59 tribes and 404 genera and 15162 species have been reported till date from the world (Holldobler and Wilson, 1990; www.antwiki.org, 2015). The most specious sub-families of ants refer to Myrmicinae (4377 species, 155 genera), Formicinae (2458 species, 49 genera), Ponerinae (1299 species, 42 genera), Dolichoderinae (554 species, 22 genera) and Pseudomyrmecinae (197 species, 3 genera) (Bolton, 1995).

The review of literature indicates that several authors like Linnaeus (1758,1764), Fabricius (1787,1798), Latreille (1802), Guerin (1844), Mayr (1861), Girard (1879), Emery (1901,1906,1910,1911,1912,1921), Wheeler (1913), Chapman (1951), Bolton (1994,1995) etc have published their valuable work on ants.

Formicidae of World

Schuckard (1840) published a monograph of Dorylidae, the family of Hymenoptera. Nylander (1846) added notations in monograph of *Formicarum borealium* of European origin. Smith (1853) studied the genus *Cryptocerus* of family Myrmicidae. Smith (1857) provided catalogue of hymenopterous insects collected at Sarawak, Borneo and Malcca and Singapore including ants. He also provided catalogue of hymenopterous insects including ants in the collection of British Museum (Smith, 1858). Smith (1877) described new species of the genera *Pseudomyrma* and *Tetraponera* belonging to the family Myrmicidae. Emery (1901) added notes on sub-families Dorylinae and Ponerinae. Ashmead (1905a) studied new Hymenoptera from the Philippines including ants. Ashmead (1905b) provided a skeleton of a new arrangement of the families, sub-families, tribes and genera of the ants. Ashmead (1906) classified the foraging and driver ants, of family Dorylidae and described the genus *Ctenopyga* Ashm. Emery (1906) also added notes on *Prenolepis vividula.*

Emery (1910, 1911, 1912, 1921) studied sub-families of the family Formicidae *viz.*, Dorylinae, Ponerinae, Dolichoderinae and Myrmicinae. Arnold (1915, 1916, 1917, 1920a, 1920b, 1922, 1924, 1926) published monographs on the Formicidae of South Africa. Brown (1958) reviewed the ants of New Zealand. Kannowski (1957) made a note on the ant *Leptothorax provancheri* Emery. Brown and Kempf (1969) described three new species of the genus *Acanthognathus.* Hung (1970) revised ants of the subgenus *Polyrhachis* Fr. Smith. Bolton (1971) reported two new sub arboreal species of the ant genus *Strumigenys* from West Africa. Bolton (1972) recorded two new species of the ant genus *Epitritus* from Ghana, with a key to the world species. Beatson (1972) studied the Pharaoh's ants as pathogen vectors in hospitals. Bolton (1973a) studied the ant genera of West Africa and a synonymic synopsis with keys. Bolton (1973b) reported the ant genus *Polyrhachis* Fr. Smith in the Ethiopian region. Bolton (1973c) studied a remarkable new arboreal ant genus from West Africa. Bolton (1974a) revised the Palaeotropical arboreal ant genus *Cataulacus* F. Smith. Bolton (1974b) recorded new synonymy and a new name in the ant genus *Polyrhachis* Fr. Smith. Bolton (1974c) revised the ponerine ant genus *Plectroctena* F. Smith. Bolton and Collingwood (1975) published a handbook for the identification of British insects. Bolton (1976) studied the ant tribe Tetramoriini and reviewed smaller genera and revised *Triglyphothrix* Forel. Bolton (1977) studied the ant tribe Tetramoriini and the genus *Tetramorium* Mayr from the Oriental and Indo-Australian regions.

Aktac (1977) studied the myrmeco fauna of Turkey. Bolton (1978) exposed the tribe Tetramoriini for ants. Bolton (1979) also studied the ant tribe Tetramoriini and the genus *Tetramorium* Mayr from Malagasy region and the new world. Bolton *et al.* (1979) reported new West African ants of the genus *Plectroctena* with ecological notes. Bolton (1980) reported the genus *Tetramorium* Mayr from the Ethiopian zoogeographical region. He revised the ant genera *Meranoplus* Smith, *Dicroaspis* Emery and *Calyptomyrmex* Emery of Myrmicinae from the Ethiopian zoogeographical region (Bolton 1981a, b).

Edwards and Baker (1981) studied the distribution and importance of the Pharaoh's *Monomorium pharaonis* (L.) from National Health Service hospitals in England. Andersen (1982) worked in semiarid area of North Western Victoria and tropical savanna at Northern Territory and reported 105 and100 species of ants respectively which were richest ant fauna of the world. Bolton (1982) described Afrotropical species of the Myrmicinae ant genera *Cardiocondyla, Leptothorax,*

Melissotarsus, Messor and *Cataulacus* and reviewed the *Solenopsis* genus - group and the Afro tropical genus *Monomorium* Mayr (Bolton, 1987). Bolton (1983) studied the Afrotropical Dacetine ants. Bolton (1984) also added on diagnosis and relationships of the Myrmicinae ant genus *Ishakidris* gen.n. In 1986a he provided a taxonomic and biological review of the ant genus *Rhoptromyrmex.* Bolton (1986b) studied the apterous females and shift of dispersal strategy in the *Monomorium salomonis* group. Wilson (1987) reported the arboreal ant fauna of Peruvian Amazon forest. Bolton (1987) reviewed the *Solenopsis* genus-group and the genus *Monomorium* Mayr. Bolton (1988) also reviewed *Paratopula* Wheeler, a genus of Myrmicinae ants. Bolton and Marsh (1989) studied the Afrotropical thermophilic ant genus *Ocymyrmex.* Agosti and Bolton (1990) reported the new characters to differentiate the ant genera *Lasius* F. and *Formica* L. Majer (1990) studied the abundance and diversity of arboreal ants from Northern Australia.

Bolton (1991) reported new Myrmicinae ant genera from the Oriental region. Similarly, Agosti (1991) revised the oriental ant genus *Cladomyrma* with an outline of the higher classification of the Formicinae. Daniels (1991) suggested the ants were biological indicators of environmental changes. Alpert (1992) recorded the genus *Terataner* from Madagascar, while, Bolton (1992) reviewed the genus *Recurvidris* with a new name for *Trigonogaster* Forel. Bolton and Belshaw (1993) studied the taxonomy and biology of the genus *Paedalgus.* Agosti (1994) provided the phylogeny of the ant tribe Formicini and described the new genus *Bajcaridris, the* subgenus *Iberoformica* was synonymised with *Formica,* he also provided the synopsis, diagnosis and keys to the genera. Agosti *et al.* (1994) reported 104 ant species belonging to 41 genera from Malaysia. Bueno and Fowler (1994) reported exotic and native ant fauna of Brazilian hospitals. Agosti (1995) revised the South American species of the genus *Probolomyrmex.* A taxonomic and zoogeographical census of ant taxa was made with a new general catalogue of the ants of the world by Bolton (1995a, b).

Andersen and Clay (1996) have recorded 248 species of ants from 32 genera in semiarid area of Australia. Fisher (1997) studied the biogeography and ecology of the ant fauna of Madagascar. Folgarait (1998) reviewed the biodiversity of ants and their roles in ecosystem functioning. Whiteford (1999) studied the densities of active ant colonies in three habitats: creosote bush shrub land, grassland and shinnery-oak mesquite dunes and studied diurnal foraging patterns at bait boards. Currie *et al.* (1999) studied the fungus growing ants to use antibiotic producing bacteria to control garden parasites. Lobry de Bryuyn (1999) observed the ants as bioindicators of soil function in rural environments. Bolton (1999) reported the ant genera of the tribe Dacetonini.

Callcott *et al.* (2000) studied seasonal pattern and adaptability of an isolated red imported fire ant population in Eastern Tennessee. They evaluated winter survivability of *Solenopsis invicta* and local myrmecofauna. Boursaux-Eude and Gross (2000) studied the symbiotic associations between ants and bacteria. Pitts and Pitts - Singer (2001) reported that the phorid flies of the genus *Pseudacteon* were potential biocontrol agents against the imported fire ants, *Solenopsis invicta* Buren and *S. richteri* Forel in North America. ITO *et al.* (2001) investigated the ant fauna of Bogor Botanic Garden, West Java Indonesia by using various sampling methods like collection of ant on tree trunk, by using handy sifter, pitfall traps, sugar baits; on bamboo shoots, searching for colonies, foraging workers *etc*. In all, 216 species were reported from almost all the sub-families, also described two new species of the genus *Leptanilla.* Watt *et al.* (2002) studied the diversity and abundance of ants in relation to forest disturbance and plantation establishment in Southern Cameron. Longino *et al.* (2002) studied the ant fauna of a tropical rain forest and estimated three different ways of species richness. Bolton and Brown (2002) erected new genus *Loboponera* and reviewed the Afrotropical *Plectroctena* genus group.

Bolton (2003) published a book on synopsis and classification of Formicidae. Astruc *et al.* (2004) studied the phylogeny of ants based on morphology and DNA sequence data. Radchenko (2004) revised nineteen species of the genera *Leptothorax* and *Temnothorax* distributed from Mongolia to the Pacific Ocean and provided a key to their identification. Boulton *et al.* (2005) examined the role of cattle grazing, plants and soil attributes on species richness, abundance and composition of ground - dwelling ants in Northern California serpentine and nonserpentine grasslands, also analyzed the relationship between three numerically dominant ant species, ant species richness and abundance.

Tinaut *et al.* (2005) reported a total of 15 species of Myrmicinae parasitic ants from the Iberian Penin Sula; they also reviewed on their biology, relation to parasitism, their distribution and taxonomy. Haneda *et al.* (2005) investigated effectiveness of two ant sampling methods *i.e.*, pitfall and yellow-pan traps in three forest habitats in Malaysia. Philpott *et al.* (2006) studied effects of management intensity and season on arboreal ant diversity and abundance in coffee agro ecosystems. Radchenko *et al.* (2006) described the new species and the queen of *Myrmica schoedli* sp.n. collected from the mountains of northern Vietnam.

Hughes (2006) reviewed four species of wood ants in Scotland and provided ecology, identification, distribution, conservation and recommendations for monitoring and management of these species. Eguchi *et al.* (2006) studied the Oriental species of the genus *Probolomyrmex*. Longino (2006) made a taxonomic review of the genus *Myrmelachista* in Costa Rica. Dunn *et al.* (2007) developed a global database for the diversity of ants and described the database and aspects of its limitations and possibilities. Ogata and Okido (2007) revised the genus *Perissomyrmex* including a definition of the taxon, description of a new species from China and discussed the phylogenetic position in the tribe. Berghoff and Franks (2007) recorded army ant, *Cheliomyrmex morosus,* firstly in Panama and its high associate diversity. Calcatera *et al.* (2007) conducted a survey in the same latitudes in Central Western Argentina to detect the presence of *Solenopsis* fire ants and their parasitoid flies in central Chile and central Western Argentina. Groc *et al.* (2007) studied the applicability of sampling methods *i.e.*, Winkler extractors, pitfall traps, baiting and manual collection conducted in the tropics to create ant species diversity in temperate environments. Stoll *et al.* (2007) reported the bacterial micro biota associated with ants of the genus *Tetraponera.* Bolton (2007a) studied the taxonomy of the Dolichoderinae ant genus *Technomyrmex* Mayr based on the worker caste. Bolton (2007b) also suggested how to conduct large-scale taxonomic revisions in Formicidae.

Bollazzi *et al.* (2008) studied the soil temperature, digging behavior, and the adaptive value of nest depth in South American species of *Acromyrmex* leaf-cutting ants. Lange *et al.* (2008) studied the predacious activity of ants in conventional and in no-till agriculture systems. Bolton *et al.* (2008) recorded new synonyms for Neotropical Myrmicinae ants. Bolton and Fisher (2008a) reported the Afrotropical Ponerinae ant genus *Asphinctopone* Santschi. Bolton and Fisher (2008b) studied the Afrotropical Ponerinae ant genus *Phrynoponera* Wheeler. Bolton and Fisher (2008c) also studied Afrotropical ants of the Ponerinae genera *Centromyrmex* Mayr, *Promyopias* Santschi and *Feroponera* with a revised key to genera of African Ponerinae. Holldobler and Wilson (2009) published a book on ants "The Superorganism." Neves *et al.* (2010) compared the composition and richness of arboreal ants between the dry and wet seasons in three successional stages of a tropical dry forest in Brazil. Calvo and Schultz (2010) described three new species of *Myrmicocrypta* from Brazil and Peru and discussed characters that define the genus and reviewed its phylogenetic position within the tribe Attini.

Rabeling and Bacci (2010) described a new species of workerless inquilines in the fungus - gardening ant, genus *Mycocepurus* Forel. Bolton and Fisher (2011) studied the taxonomy of Afrotropical and West Palaearctic ants of the Ponerinae genus *Hypoponera* Santschi. Lachaud and Lachaud (2012) reported diversity of species and behavior of Hymenopteran parasitoids of ants. Resende *et al.* (2013) analyzed assemblages of arboreal ants in different vegetation within the Tropical Moist Forest domain and suggested that rubber tree plantation can be a good matrix for the maintenance of some ant species. Bolton (2013) made a catalogue of the Ants of the World and very recently, Mortazavi (2015) investigated aphid-ant mutualistic associations. Francisco Hita Garcia *et al.* (2015) revised the genus *Proceratium Roger* in Fiji.

Formicidae of India

The pioneer work on taxonomy and biodiversity of Indian ants refer to Forel (1892, 1894, 1895, 1901, 1902, 1903). Saunders (1842) made a description of two ant species from northern India. Jerdon (1851) prepared a catalogue of the ants found in Southern India. Wheeler (1917), Donisthorpe (1942), Bucher (1952) and Wilson (1953) studied the ants with respect to faunatic diversity from India. Similarly, Konnowaski (1956), Galle (1972). Reddy *et al.* (1981) made a preliminary study of ant fauna of Bisthenhatti. Kamatar (1983) studied ants of Raichur district and reported the biology and behavior of fire ant *Solenopsis geminata*. Dasmann (1984) worked on taxonomy of ants of Southern India. Tiwari and Jonathan, (1986a) reported a new species of *Liomyrmex* Mayr from Andaman Islands. Tiwari and Jonathan (1986b) also reported a new species of *Metapone* Forel from Nicobar Islands. Cherret (1989) worked on taxonomy of ants of Southern India. Bingham (1903) reported 498 species under 79 genera from India, including Myanmar and Shri Lanka. Chapman and Capco (1951) recorded 2080 species, 4 from Asian subcontinent, 41 subspecies and 684 varieties of ants spread over 176 genera from India.

Ali (1991, 1992) studied ant fauna of Karnataka. Gadagkar *et al.* (1993) studied ant species richness and diversity from some selected localities of Western Ghats of

India. Tiwari (1994) discovered two new species of genus *Myrmecina* from Kerala. Tak (1995) studied the ants of Rajasthan. Tak and Rathore (1996) reported ant fauna of the Thar Desert while, Rastogi *et al.* (1997) worked on ant fauna of the Indian Institute of Science campus, they made the surve and some preliminary observations on ants. Rastogi *et al.* (1997) also made preliminary observations of foraging strategies in the ant *Myrmicaria brunnea* and *Diacamma ceylonense* in campus of the Indian Institute of Science. Sheela and Narendran (1997) discovered a new genus and a new species of Myrmecinae from India. Kumar Sunil *et al.* (1997) studied ant species richness at selected localities of Bangalore. Basu (1997) studied the seasonal and spatial pattern of ground dwelling ants in a rain forest in Western Ghats.

From Oriental region 227 genera of ants have been reported. The number of species described from Oriental region was 2480 (771 from Oriental region and 1709 from Indo – Australian region (Tiwari, 1999).Viswanathan and Narendra (1999) studied the behavior of the ant *Myrmicaria brunnea* Saunders towards pheromones. Tiwari (1999) made taxonomic studies of ants of Southern India. Viswanathan and Narendra (2000) studied the food preference in different species of ants. Viswanathan and Narendra (2000) studied the effect of urbanization on the biodiversity of ants in Bangalore. Bharti (2001) reported two new species of *Pheidole* west wood from India. Ward (2001) studied the taxonomy, phylogeny and biology of the genus *Tetraponera* from Oriental and Australian regions. Radchenko (2003) described a new species *Perissomyrmex nepalensis* sp. nov from India and provided a key to the identification of the species the genus *Perissomyrmex*. Narendra (2003) observed responses of the Asian weaver ant, *Oecophylla smaragdina* towards high quality and quantity food substances. Bharti (2003a) studied queen of the army ant *Aenictus pachycerus*. Bharti (2003b) also reported a new species *Polyrhachis punjabi* sp. n. and one more new species of *Crematogaster* from India (Bharti 2003c). Varghese (2003, 2004a) studied ants of the Indian Institute of science campus of Bangalore. Varghese *et al.* (2004) prepared a checklist of ants from the insect museum of Bangalore. Varghese (2004b) also recorded *Strumigenys emmae* (Emery) from Bangalore, Karnataka and provided a key to Indian species. Zacharias and Rajan (2004a) reported *Discothyrea sringerensis* as a new ant species from India. Zacharias and Rajan (2004b) also reported a new species *Vombisidris humbolticola* on Indian ant plant. Karmaly (2004) reported a new species and a key to species of the genus *Polyrhachis* Smith from India.

Rajagopal *et al.* (2005) studied ant diversity in some localities of Sattur Talukas of TamilNadu. Haneda *et al.* (2005) provided two ant sampling methods for collection of ants for tropical rainforest. Ghosh *et al.* (2005) worked on ants of Rabindra Sarovar of Kolkata. Karmaly and Narendran (2006) reported Indian ants of the genus *Camponotus*. Varghese (2006) described a new species of *Dilobocondyla bangalorica* from India and also described worker, queen, male and larvae along with nesting behavior. Varghese (2006) made a description of a new species of the Ponerinae of the genus, *Emeryopone* from Karnataka. Anto and Thomas K (2007) studied the diversity of litter ant assemblage in the Wayanad region of Western Ghats from Kerala. Narendra and Kumar (2007) published a handbook of the ants of Peninsular India. Bharti (2008) redescribed the species *Crematogaster subnuda subnuda*. Savitha *et al.* (2008) studied ant diversity and distribution of ants across an urban gradient in Bangalore. Kumar and Mishra (2008) studied ant biodiversity from the Vadodara district of Gujarat and assessed density change in urban and agricultural ecosystems. Sheela (2008) published a handbook on hymenoptera including ants. Sabu *et al.* (2008) studied diversity of forest litter-inhabiting ants from Wayanad region of the Western Ghats. Vishnudas (2008) studied the woodpeckers and ants in India's shade coffee. Varghese (2009) reviewed many sub-families, tribes and genera of ants found in India.

Bharti *et al.* (2009) analyzed seasonal patterns of ants in five seasons in Punjab Shivalik range of North-West Himalaya; they analyzed 40 species for seasonal patterns and he concluded species richness. Jaitrong *et al.* (2010) studied the army ant *Aenictus wroughtonii* and related species in the Oriental region and made a description of two new species. Ramesh *et al.* (2010) studied an ant faunal diversity, species composition and effect of microhabitat from Kalpakkam, South India during dry season. Chavan and Pawar (2011) studied diversity and distribution and species richness of ant in three different habitats *viz.,* forest, grass land and human habitat from Amravati city. Sonune and Chavan (2011) studied the diversity and species richness of ants in Aurangabad district. Bharti (2011) prepared a list of Indian ants. Bharti and Wachkoo (2011) reported *Amblyopone boltoni,* a new ant species from India. Nagariya and Pawar (2012) studied the diversity and distribution of ants in Pohara forest area of Amravati region and observed three species namely Red imported fire ant, *Solenopsis invicta,* carpenter ant, *Camponotus* and Pharaoh ant, *Monomorium pharaonis,* as prominent members of the region.

Bharti *et al.* (2012) investigated two remarkable new species of *Aenictus* from India. Jaitrong and Yamane (2012) reviewed the Southeast Asian species of *Aenictus javanus* and *Aenictus philippinensis* species groups. Sonune and Chavan (2012) studied taxonomy of ants around Gautala Autramghat Santuary Aurangabad. ZSI (2012) published the fauna of ants of Maharashtra. Bharti (2012a) reported *Myrmica nefaria* sp. n. from Himalaya. Bharti (2012b) reported two new species of the genus *Myrmica* from the Himalaya. Bharti *et al.* (2012) also studied three new species of genus *Temnothorax* from Himalayas with a revised key to the Indian species. Bharti and Kumar (2012a) studied *Lophomyrmex terraceensis,* as a new ant species of the bedoti group with a revised key. Bharti and Kumar (2012b) also studied taxonomy of genus *Tetramorium* Mayr, with report of two new species and three new records including a tramp species from India with a revised key. Bharti and Kumar (2012c) reported a new species of the genus *Leptanilla* with a key to Oriental species.

Khot *et al.* (2013) studied ant diversity in an urban garden at Mumbai, Maharashtra and represented 28 species of ants from 6 sub-families. Bharti and Ali, (2013b) described a new species of the genus *Lordomyrma* from silent Valley National Park, Kerala India. Bharti and Akbar (2013a) studied the taxonomy of the genus *Strumigenys* Smith, also reported two new species and recorded five new species from India. Bharti and Akbar (2013b) reported new species of the genus *Lordomyrma* from India. Bharti and Akbar (2013c) studied the ant genus *Cerapachys* Smith from India. Kharbani *et al.* (2013) studied the seasonality of ants in terms of species diversity, foraging activity and species composition under the influence of various ecological factors with annual seasonal cycle. Kataria and Kumar (2013) studied the ant aphid association and its relationship with various host plants in agro ecosystems of Vadodara in Gujarat and noted the damage caused by aphids and the role of ants in spreading pest from one crop to another in agro ecosystems.

Sonune and Chavan (2013) studied taxonomy and diversity of the genus *Tetraponera* (Pseudomyrmicinae) from forest area of Aurangabad district. Cunha *et al.* (2013) recorded diversity and distribution of ant fauna in Hejamadi Kodi spit, Udupi district of Karnataka, India. Patkar *et al.* (2013a) studied the ants of Solapur city. They also studied the efficiency of ant collection methods in and around Great Indian Bustard Wildlife Sanctuary (Patkar *et al.*, 2013b). Bharti and Wachkoo (2014a) described a new species of carpenter ant and illustrated on the worker and gyne castes *Camponotus parabarbata* sp.n, they also provided identification key. Bharti and Wachkoo (2014b) published new synonymy of *Proceratium williamsi* Tiwari. Bharti and Akbar (2014) described new species *Meranoplus periyarensis* from India. Selvarani *et al.* (2014) studied the seasonal dispersal of leaf litter ants in Megamalai Western Ghats of India. Patkar *et al.* (2014) studied the taxonomy of ants from great Indian Bustard Wildlife Sanctuary.

Recently, Bhoje *et al.* (2014) studied the biodiversity of ants of Amba reserve forest of Western Ghats. Similarly, Kurane *et al.* (2014) studied diversity of ants from Kolhapur district of Maharashtra. Very recently, Kurane *et al.* (2015) studied the diversity and economic importance of ants of Kolhapur city. Review of literature indicates that little attension is paid on the species diversity of ants from Maharashtra although this group has tremendous economic importance.

3

Collection and Preservation Methods

Materials and methods play a very crucial role in investigating processes in living things. Therefore, choice and selection of materials and methods have tremendous importance in doing research in biological sciences. Any minor change can have drastic change in the results both desirable and undesirable. Following materials were used in completion of the research work.

A) Materials

1) Specimen Bottles (Figure 3)

Specimen bottles of size 2.5×6 cm and 2×6 cm (diameter and length) were used for preservation of ants.

2) Camel Hair Brushes (Figure 4)

Camel hair brushes no. 4 and 8 have been used for collection of ants and cleaning the preserved specimens.

3) Forceps (Figure 5)

Forceps were used for collection, carding the ants and for handling preserved insects.

4) Oven (Figure 6)

Oven of size 4× 3 feet (height and width) have been used for drying ants.

5) Insect Storage Boxes (Figure 7)

Insect storage boxes of size 30× 45 cm were used for keeping dried ants.

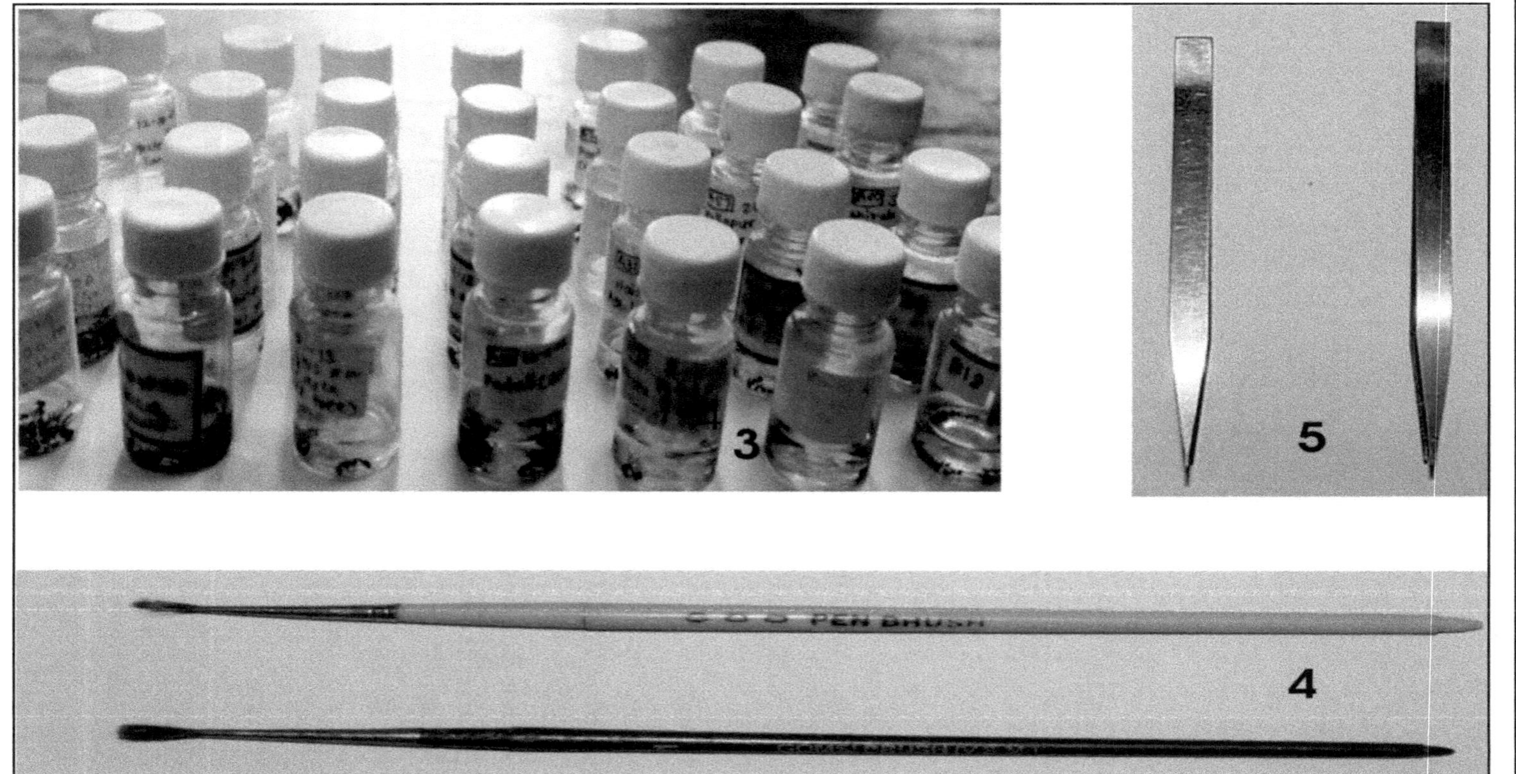

Plate 1: Figure 3: Specimen Bottles; Figure 4: Camel Hair Brushes; Figure 5: Forceps.

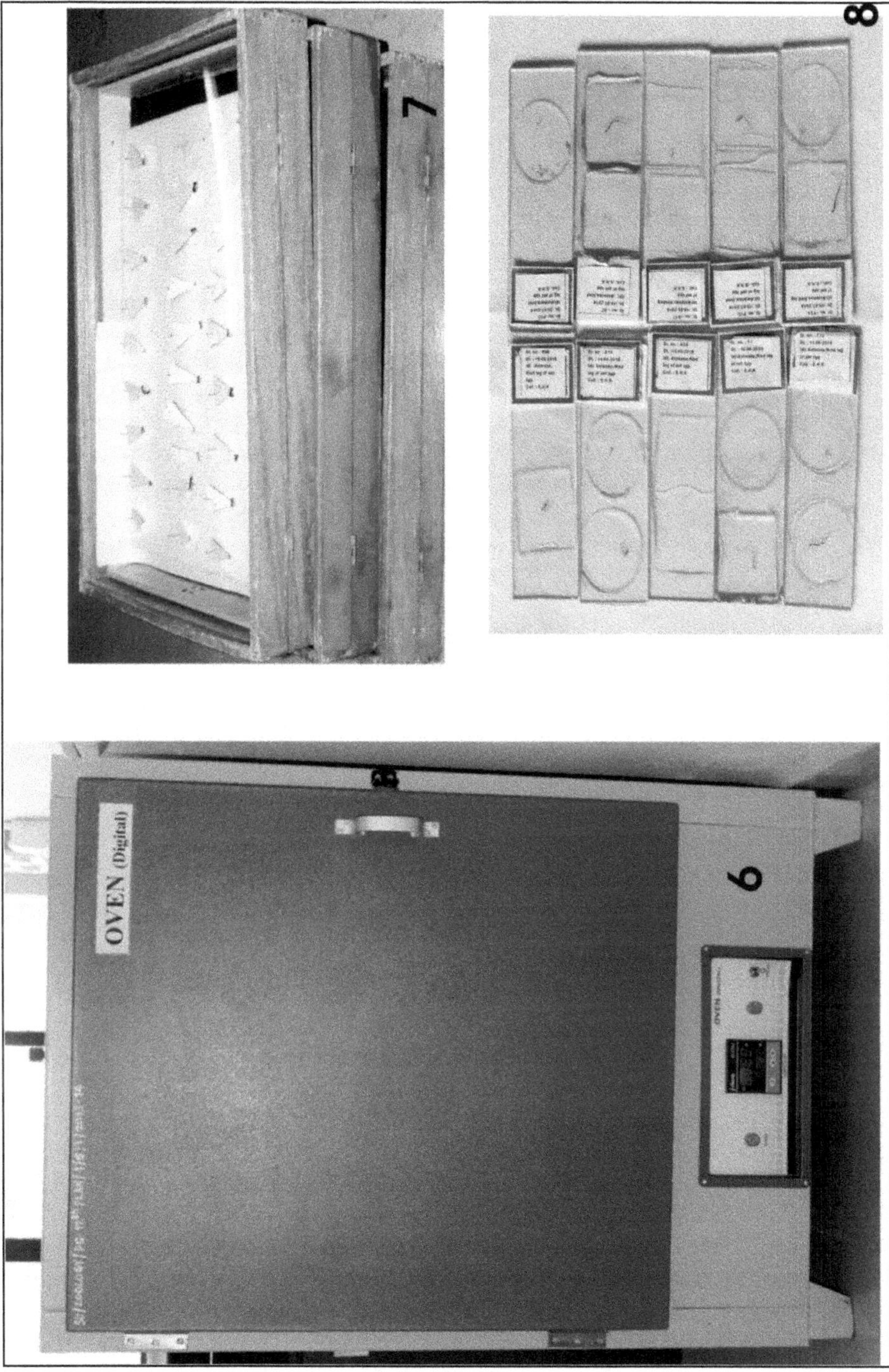

Plate 2: Figure 6: Oven; Figure 7: Insect Storage Boxes. Figure 8: Slides.

6) Slides and Cover Slips (Figure 8)

Slides and cover slips were used for preserving body parts of ants such as antenna, hind leg *etc.*

7) Compound Microscope (Figure 9)

Compound microscope with 10X and 45 X objects was used for morphological and taxonomical studies of ants.

8) Stereomicroscope (Figure 10)

Stereomicroscope (SZX-7) within build LED light source, 10X eye piece,1X objective lence and 45° binocular observation tube was used for examination of body parts and morphological characters of ants.

9) Camera (Figure 11)

Photographs of whole ant body and other various body parts have been taken by using Photographic camera, Canon IXUS 220 HS (21.1 mega pixels).

10) Card Sheet Paper

The white coloured card sheet papers were used for preserving ants.

11) Chemicals

Following chemicals were used for preservation of insects.

a) 70 per cent Alcohol for wet preservation.
b) Alcohol grades (30 per cent, 50 per cent, 70 per cent, 90 per cent and absolute) for dehydration of insects.
c) Xylene- cleaning the specimens and slides.
d) D.P.X. for mounting the insects and body parts of ants on slide.
e) Glue for carding the specimens.
f) Naphthalene balls for avoiding fungal infection and damage of insects.

B) Methods

Ants were preserved by following methods:

1) Dry Preservation Method

The ants were collected from agro, horticultural and forest ecosystems from Western Maharashtra specially Kolhapur, Sangli and Satara. The ants were collected with the help of camel hair brush and forceps by using hand picking method. Later, carding of ant species were made in scientific manner. Ants were glued on small triangular strip of paper between the middle and hind coxa. Then pushed the pin to the base of card and labeled. The specimens were dried in oven at 60°C temperature. Slides were prepared for morphological studies. Dried specimens were preserved in insect storage boxes. Naphthalene balls were used for avoiding fungal infection in the boxes. Morphological/taxonomical studies of ants were made by using compound microscope/stereo –microscope. Later, photography was made by using

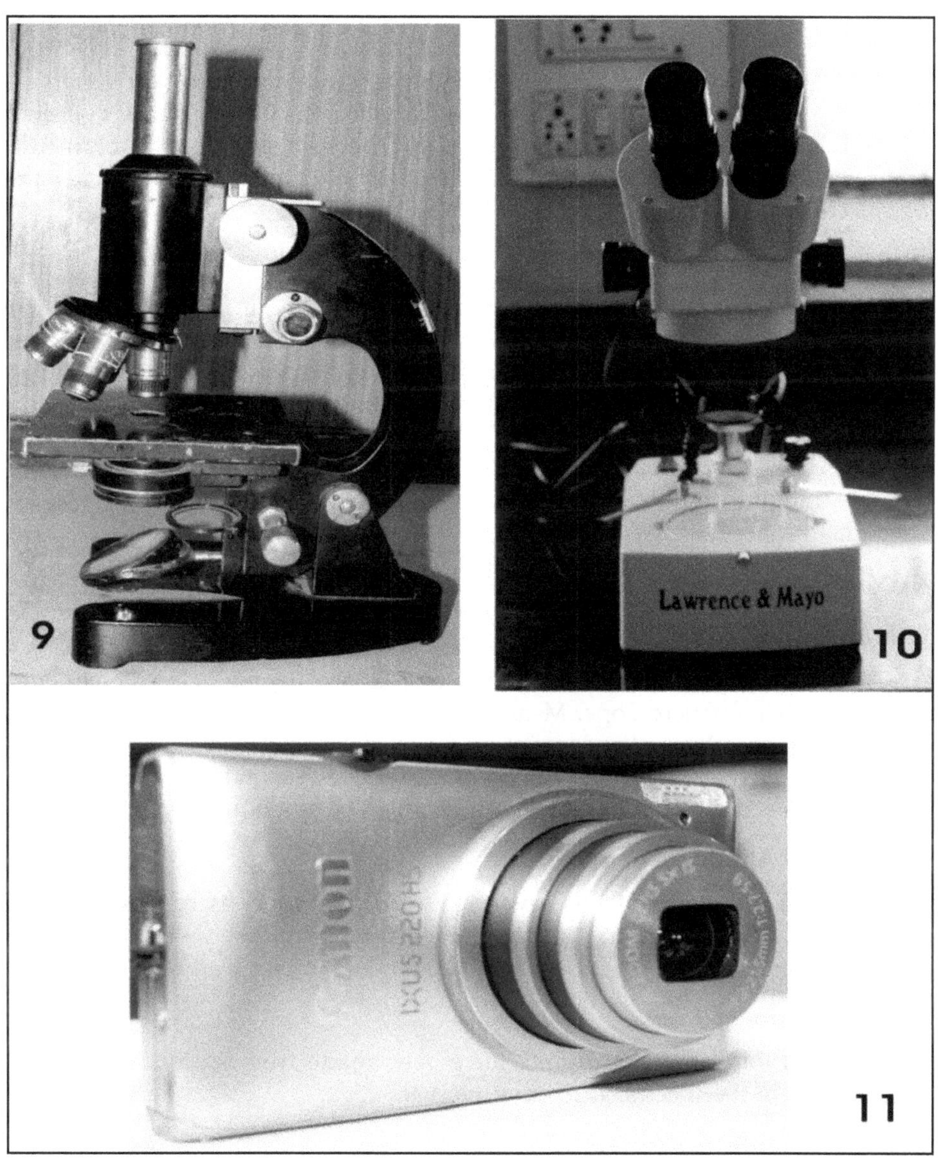

Plate 3: Figure 9: Comopound Microscope; Figure 11: Steriomicroscope; Figure 11: Camera.

photographic Canon IXUS 220 HS camera with 12.1 X optical zoom. Measurements of body parts of species were taken in mm in the description. Colour photographs of whole ants and their body parts were made by using Stereomicroscope which was used for description.

2) Wet Preservation Method

Ant species were collected from agro, horticultural and forest ecosystems of study area with the help of camel hair brush and forceps by using hand collection methods. The collected ants were preserved in 70 per cent alcohol. Specimen bottles were labeled with details of location, date of collection, time of collection and name of collector.

3) Slide Preparation Method

Preserved ants were boiled in 10 per cent KOH for 5 min. Then the specimens were washed in tap water to remove the KOH and again washed with distilled water. Then dehydration was done by using alcohol grades (30 per cent, 50 per cent, 70 per cent, 90 per cent, and 100 per cent) and then rinsed in Xylene. Transparent specimens were kept on slide and mounted in D.P.X. Then holding the forcep by fingers, the cover slip bring in contact with slide and slowly lowered on the mounting medium to avoid trapping air bubble in the mountant. The slides were kept horizontal and used after proper drying (1 week).

4) Method of Study

a) Morphology

Taxonomical studies have been made on ants with the help of compound microscope, stereomicroscope. Measurements of species and body parts of ants were taken in mm during the description. Coloured photographs of ants have been taken as supportive part of description of the species. Taxonomical description of the ants has been made as per the terminology given by Bolton (2013).

b) Seasonal Abundance

Seasonal abundance of ants have been studied by spot observation of species from different study spots at 15 days intervals throughout the year during course of study by one man one hour search method. Few samples were also collected.

c) Distribution of Ants

Distributional records of ants have been made from study area by spot observations of the species, at 15 days interval.

b) Host Records

Host records of ants have been prepared by observing the ants on specific plants in the study area. Later, the flora and fauna have been identified by consulting appropriate literature.

Study Area

The study area shown in Figure 12 is from three districts of Western Maharashtra *viz.*, Kolhapur, Sangli and Satara.

Kolhapur district (Figure 13) has an area of 7,585 sq. km. It lies between 15 °43′ to 17°17′ North latitude and 73°40′ to 78°42′ East longitude. Total forest coverage is about 1672 sq.km. Out of which 563 sq.km forests is reserved forest and 417sq.km is a protected forest and total forest area is about 22 per cent.

In Kolhapur district subtropical evergreen, moist deciduous and semi – evergreen and dry deciduous forests are found. In the subtropical evergreen forest, the principal trees found are Sag, Hirada, Jambhul, Amba, Fanas, Behada, Kinjal etc. In semi-evergreen and moist deciduous forest the flora refers Arjun, Mango, Shesum, Bhava, Ain, Umber, Biba etc. Dry deciduous forests have been classified as reserve and protected forest, known for the firewood and grass are the main marketable products from the forest.

The overall climate of Kolhapur district is salubrious and all the seasons in Kolhapur district are moderate. In summer temperature rises up to 41.68°C in month April and decreases to 14.44°C during the month December. The average humidity of district is 58 per cent. In the district climate is temperate in plains and cool in Western Ghats. The Eastern regions represent dry weather and it experiences hot winds during April and May. The Kolhapur district receives its major rainfall (Figure 14) from the South west on an average of 1600 mm. The rainy season is from June to October and rainfall varies from about 600 mm in Shirol tahsil and 5000 mm in Gaganbawda tahsil.

Kolhapur district has 12 tahasils *viz.*, Ajara, Radhanagari, Gaganbawda, Chandgad, Gadhingalaj, Bhudergad and Shahuwadi comes under hilly region of Western Ghats. Karveer, Panhala, Kagal, Shirol and Hatkanangale come under Eastern part of district. In the district there are main six rivers these are Krishna, Wrana, Panchganga, Dudhgahga, Vedaganga and Hiranyakeshi which flows east through Western Ghats.

In present study Amba Ghat, Radhanagari, Gadhingalaj, Ajara, Chandgad, Gaganbawda have been selected for ant diversity. The chief crops in Kolhapur district are Wari, Nachani, Sawa, Ratala, Rice, Jawar, Groundnut, and Wheat where as Sugarcane is highly cultivated. Vegetables like brinjal, tomato, cabbage and chili are cultivated. Fruit crops like mango, coconut, guava, jambhul, cashew nut, Curcuma, binger and tobacco are also important crops cultivated and some medicinal plants also cultivated.

Sangli district (Figure 13) is situated between 16°45′ and 17°30′ North latitudes and 73° 42′ and 75° 40′ East longitude. Sangli district bounded on East by Bijapur district of Karnataka State, West by Ratnagiri and South by Kolhapur (MS) and Belgum district of Karnataka State, Satara and Solapur lies on North boundaries of the district. Sangli district falls partly in Krishna basin and partly in Bhima basin. An eastern drought porne area is represented by Miraj and Tasgaon tahasils; North Eastern part is Khanapur and Atpadi, Kavate Mahankal and Jat tahsils. Western hilly

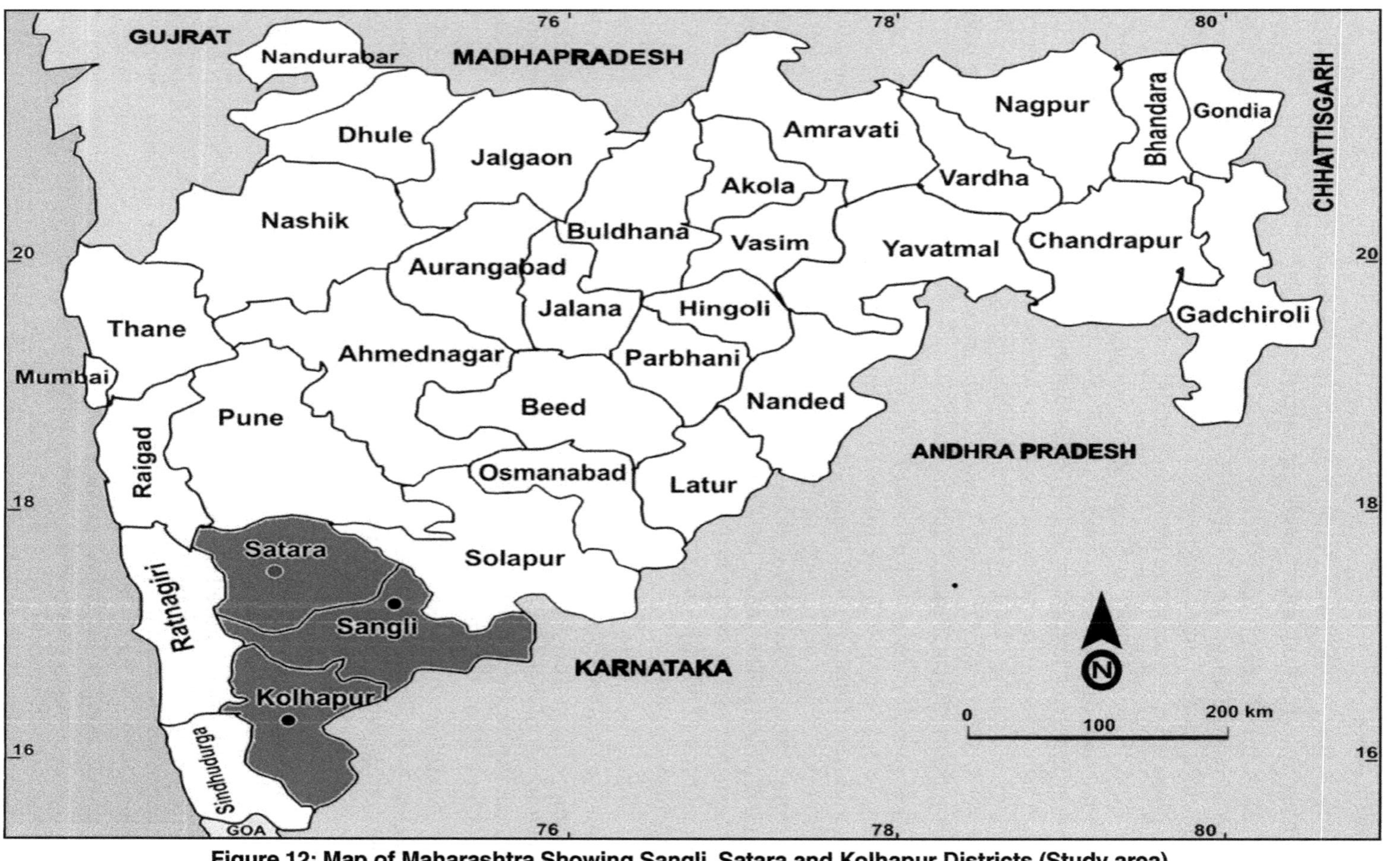

Figure 12: Map of Maharashtra Showing Sangli, Satara and Kolhapur Districts (Study area).

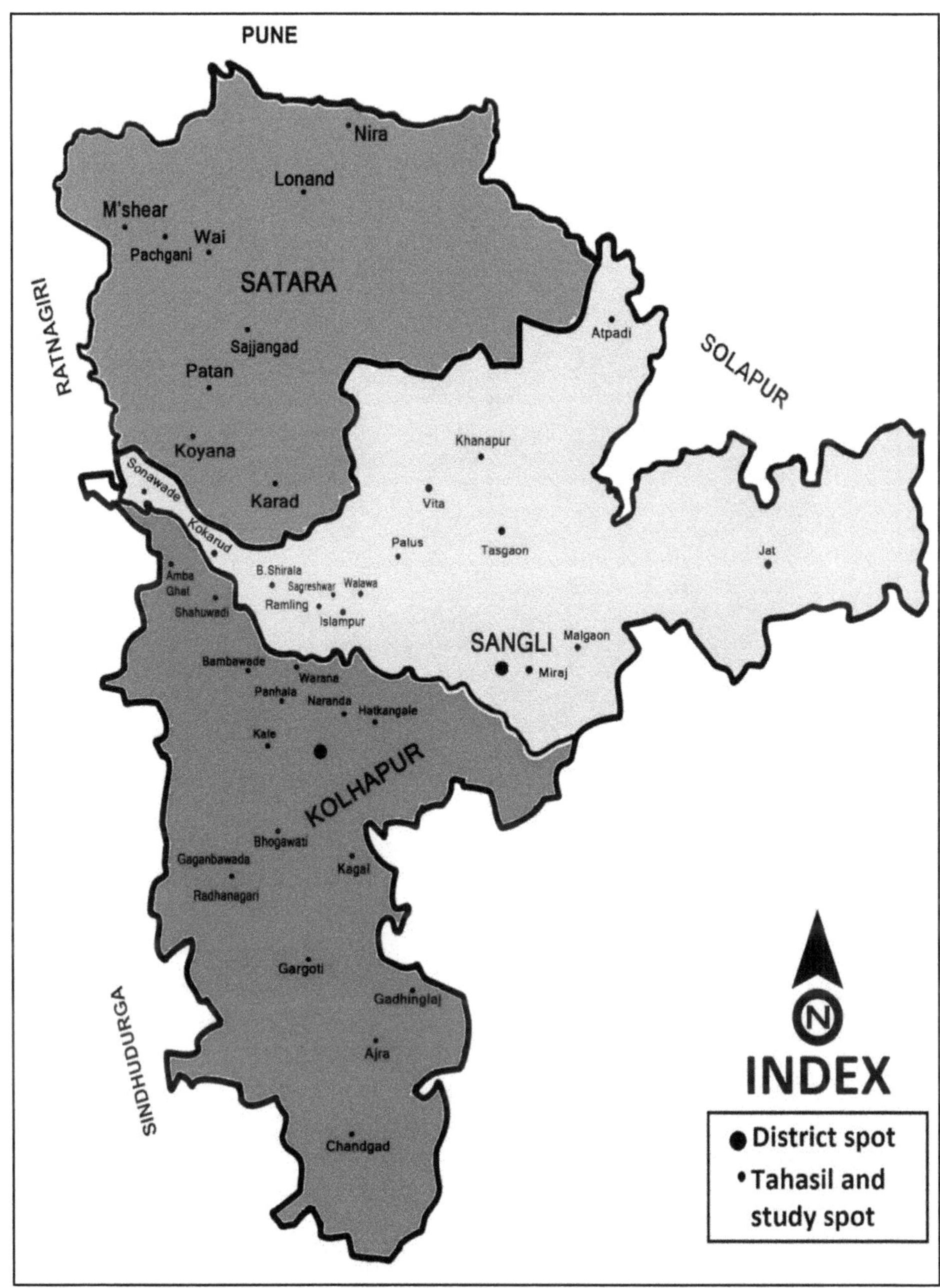

Figure 13: Map of Kolhapur, Sangli and Satara Districts showing Collection Spots.

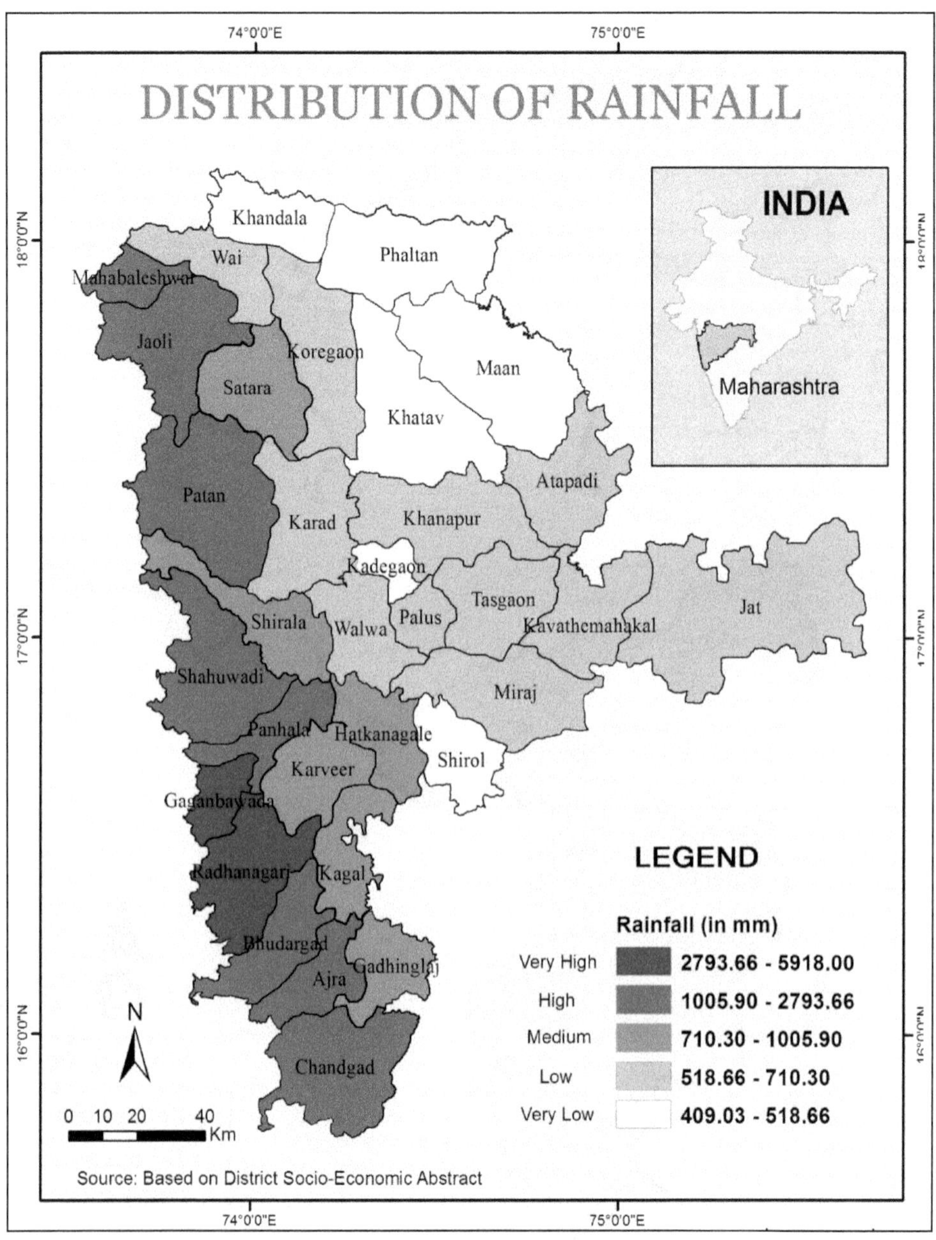

Figure 14: Average Annual Rainfall of Kolhapur, Sangali and Satara Districts (2012-2014).

Plate 4: Study Spots. Figure 15: Dam of Gaganbawada; Figure 16: Forest of Gaganbawada; Figure 17: Study Spot of Sagareshwar; Figure 18: Valleys of Western Ghats (Gadhingalaj).

Plate 5: Figure 19: Forest of Chandali; Figure 20: Hilly Region of Koyana; Figure 21: Study Spot of Gadhingalaj; Figure 22: Forest in Amba Region.

area of Shirala tahsil with heavy rainfall (Figure 14). The climate of Sangli district is hotter and drier towards the east and humid towards the west with temperature range between 10.3°C in July to 41.5°C The study spots in Sangli district refers to Sangli, Miraj, Tasgaon, Jat, Vita, Sagreshwar, Islampur, Battis Shirala have been selected for ants diversity.

The district Satara (Figure 13) occupied 10,492 sq.km and which lies between 17° 5′ latitude and 18° 1′ N and longitude 73°33′ and 74°74′ E. It is bounded by Pune at northern side and Sangli at southern side. Satara district contain eleven tahasils namely Satara, Patan, Medha, Wai and Mahabaleshwar, Jawali which comes under hilly region. Maan, Khatav, Phaltan, Koregaon, Khandala and Karad come under plain region. Koyana dam and Koyana Wild Life Sanctuary are important features of Satara. In present study Patan, Satara, Ghatmahtha, Mahabaleshwar, Karad, Wai, Pachgani have been selected for ant diversity studies.

The important crops cultivated in Satara districts refer to sugarcane, wheat, paddy, soyabean, groundnut, red gram and several vegetables and horticultural crops.

4

Biodiversity of Ants

Introduction

Biodiversity is the degree of variation of life within ecosystem; it may be a morphological, anatomical, physiological, embryological, cytological, genetic and molecular. Biodiversity studies are essential for understanding flora and fauna with respect to their occurrence, abundance and utility for the betterment of humans and environment. In terms of numbers, dominance and ecological importance, Hymenoptera is fundamentally significant order (LaSalle and Gauld, 1993). The family Formicidae alone of this order can be used to exemplify the importance of insects and biodiversity (Holldobler and Wilson 1990, Agosti *et al.*, 2000, LaSalle and Gauld 1993). LaSalle and Gauld (1993) noted that ants strongly influence the functionality of ecosystems. Ants dominate by their massive biomass, manipulate species composition, influence trophic interactions, possess numerous mutualistic and symbiotic relationships, and shape both the abiotic (*e.g.* moving soil) and biotic (*e.g.* plant-ant mosaics) matrix of community interactions.

Ants are everywhere, but run much of the terrestrial world as the premier soil turners, channelers of energy and dominatrices of the insect fauna. They employ the most complex forms of chemical communication of any animals and their social organization provides an illuminating contrast to that of human beings (Holldobler and Wilson, 1990). Since their origin in Cretaceous period ants have undergone enormous diversification. Ants belong to the family Formicidae which is largest of order Hymenoptera. The family Formicidae contain 26 sub-families with 14,711 valid species and 428 valid genera (Bolton, 2011), out of these 152 species are listed by IUCN. From India, 12 sub-families are represented by 87 genera with 652 species (Bharti, 2011). They have direct impact on ecosystem, tropic interactions, their biomass and energy consumption is greater than all the vertebrate fauna combined (Holldobler and Wilson, 1990, 2009).

Ants participate in symbioses, both facultative and obligate, with more than 465 plant species in over 52 families (Jolivet, 1996), with thousands of arthropod species (Kistner,1982; Holldobler and Wilson, 1990 and 2009) and with as-yet unknown numbers of fungi and microorganisms (Schultz and McGlynn, 2000; Mueller *et al.*, 2001; Holldobler and Wilson, 2009 and Lach *et al.*,2010).

Ants display remarkable adaptive strategies and specializations. Among insects, ants have emerged as one of the useful and effective bio-indicators due to this reasons *i.e.* they were diverse (with more than 14000 species) and were found abundantly in almost every terrestrial habitat in the world.

Materials and Methods

The ants were collected from agro, horticultural and forest ecosystems from Western Maharashtra specially Kolhapur, Sangli and Satara at 15 days interval by one man one hour search method. During the course of study 2012-2015. Ants have been collected and preserved as per the procedure given in the chapter materials and methods. Morphological diversity of ants has been studied with the help of compound microscope by consulting appropriate literature (Bingham, 1903, Bolton, 1994). Body parts such as head, thorax, abdomen and their appendages have been taken into account for morphological studies.

Morphological Consideration

The body of ant (Figure 23) is divided into head, thorax and abdomen.

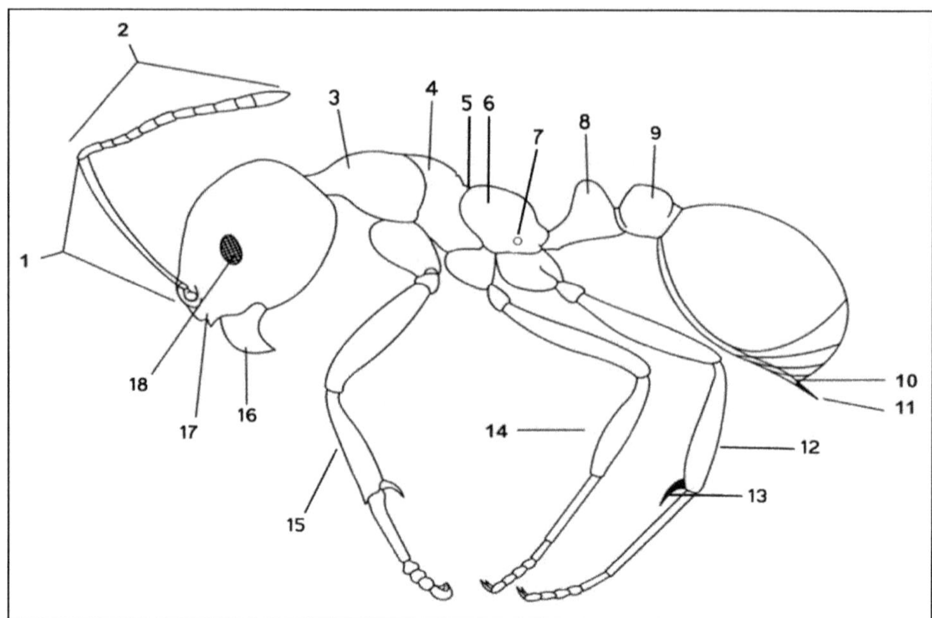

Figure 23: Morphological Features of Ants.

1. Scape; 2. Flagellum; 3. Pronotum; 4. Mesonotum; 5. Metanotum; 6. Propodeum; 7. Metapleuran gland; 8. Petiole; 9. Post petiol; 10. Pygidium; 11. Sting; 12. Hind leg; 13. Tibial spur; 14. Mid leg; 15. Fore leg; 16. Mandibles; 17. Clypeus; 18. Eye.

Head (Figure 24)

Head of ant (Figure 24) varies enormously in size and shape. It is concave or rounded, hard, heavily sclerotized, smooth or sculptured. The colour of head is also varies *i.e.* yellow, brown, black, red, metallic green, blue covered with setae or hairs. Three ocelli are rarely present in worker. The head contains eyes, clypeus, frons, vertex, genae, mouth parts and antennae. The size and the position of eyes are variable. Eyes placed downward on the head or placed upward on the head. Compound eyes well developed but in some species eyes are absent.

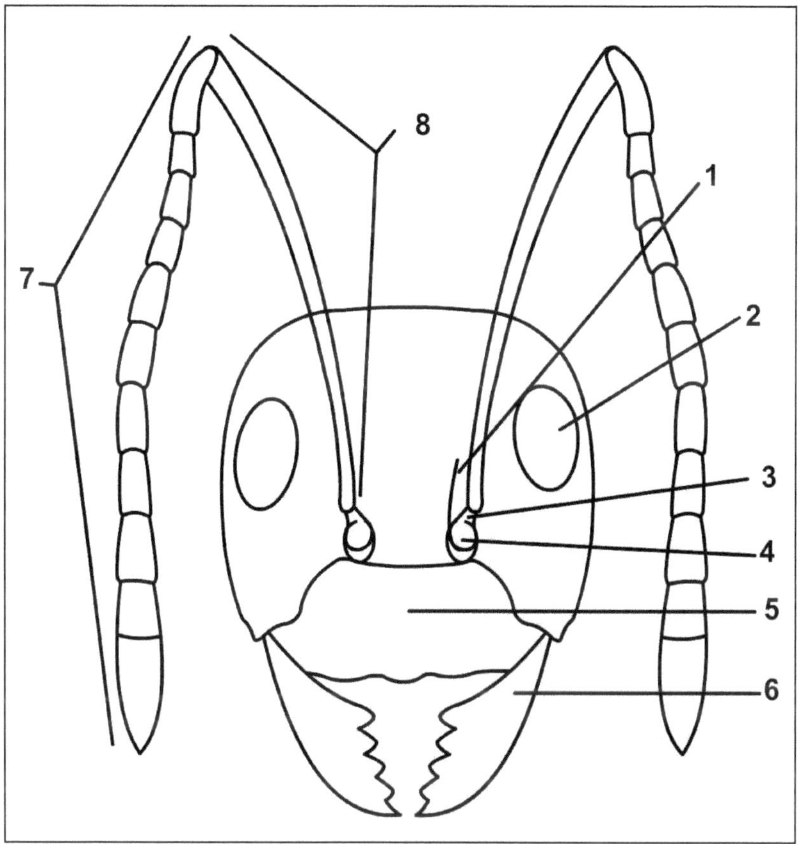

Figure 24: Morphological Features of Ant Head.

1. Frontal carina; 2. Eye; 3. Torulus; 4. Antennal scrobe; 5. Clypeus; 6. Mandible; 7. Flagellum; 8. Scape.

The mouth parts consist of mandibles, maxillae, labium and labrum. The clypeus and labrum are distinct. Clypeus broad or narrow consists of short hairs or long hairs. The region bounded anteriorly by posterior end of clypeus is the frons. It is a small, triangular area. The shape, size of frontal area is variable and very important. The occiput present between the vertex and foramen. The portion present anterior to the eyes and lateral to the frontal carinae refer to genae. Frontal carinae are a pair of longitudinal ridges on the head. In some groups frontal carinae are absent. Mandibles are variable and articulated on the lateral corners of the head and are seen below the posterior margin of the clypeus. The numbers of segments on the maxillary and labial palps have an importance in identification of species.

Antennae vary greatly with species and are very important. Antennae are geniculate (Figure 25) with 4-12 segments in workers. Scape is first antennal segment articulated from antennal socket. Scape is broad and longer followed by 3-11 short segments which constitute pedicel and flagellum. The scape and flagellum meet at an angle so that antenna appears bent (geniculate).

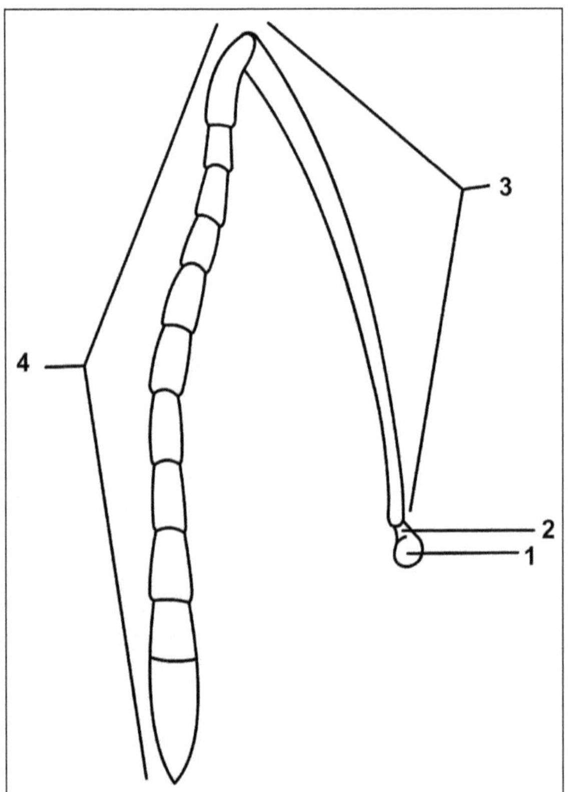

Figure 25: Morphological Features of Antenna of Ant. 1.Torulus; 2. Antennal scrobe; 3. Scape; 4. Flagellum.

Thorax (Figure 23)

In ants the thorax is divided into prothorax, mesothorax and metathorax which are leg bearing body segments. The lateral sclerites of the thorax are propleuron, mesopleuron and metapleourn. The metapleuron consist of a gland called as metapleuron gland. But in all ants, addition to these three segments, the tergite of the first abdominal segment (propodeum) is fused to the thorax region. True thorax + propodeum is also named as alitrunk or mesosoma. The thorax is broader than head. It is variable in shape, size and appearance with sex, castes and with different species.

Legs (Figure 23)

The legs are well developed; three pairs of legs are present *i.e.* fore leg, mid leg, hind leg. Fore legs and mid legs are modified for walking and hind legs modified for running, jumping etc. Hind legs (Figure 26) are longer than body. The legs of ants consist of coxa which is first basal segment of any leg and articulates within coxal cavity in ventral thorax. Trochanter is small and present between coxa and femur. Femur is the single jointed and longest segment. Tibia having spur, may or may not be present in all legs which is single or double, pectinate. Tarsi with -5-segments the apical joint armed with two claws which may be pectinate and simple.

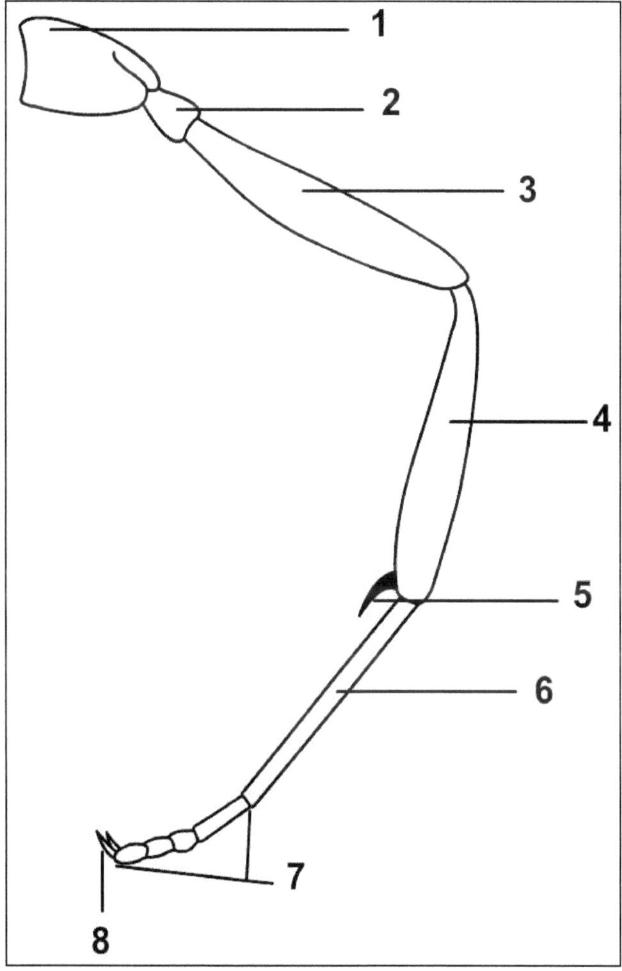

Figure 26: Morphological Features of Hind Leg of Ant.

1. Coxa; 2. Trochanter; 3. Femur; 4. Tibia; 5. Tibial spur; 6. Tarsus; 7. Pre-tarsus; 8. Tarsal claws.

Abdomen (Figure 23)

Abdomen consists of 7 segments. The first abdominal segment is fused with thorax called as propodeum. The second segment is reduced and separated from the remaining abdominal segments and forms a waist of 1 or 2 nodes and forms the petiole. The third segment is also reduced and separated to form the post-petiole respectively.This petiole and post-petiole is separated from abdominal segment by narrow constriction. The rest of the segments constitute the gaster (abdominal segment 4 to 7). It is broad and full sized. This segment consists of a pair of sclerites, dorsal tergite and ventral sternite. In workers last abdominal tergite is called as pygidium and last visible sternite is called as hypopygium.

Order Hymenoptera is divided in two suborders, twenty super families and ninety one families (http:/www.antwiki.org/).

Classification of Ants

Suborder: Symphyta

Super Family	:	Cephoidea
Family	:	Cephidae
Super Family	:	Megalodontoidea
Family	:	Megalodontoidae
Super Family	:	Orussoidea
Family	:	Orussoidae
Super Family	:	Siricoidea
Families	:	Anaxyelidae
	:	Siricidae
Super Family	:	Tenthredinoidea
Families	:	Argidae
	:	Blasticotomidae
	:	Cimbicidae
	:	Diprionidae
	:	Pergidae
	:	Tenthredinidae
Super Family	:	Xyelodiea
Families	:	Xylidae
	:	Xyphydriidae

Suborder: Apocrita

Infraorder: Aculeata

Super Family	:	Apoidea
Families	:	Ampulicidae
	:	Andrenidae
	:	Apidae
	:	Colletidae
	:	Crabronidae
	:	Dasypodaidae
	:	Halictidae
	:	Heterogynaidae
	:	Megachilidae
	:	Meganomidae
	:	Melittidae
	:	Stenotritidae
	:	Sphecidae
Super Family	:	Chrysidoidea
Families	:	Bethylidae
	:	Chrysididae
	:	Dryinidae
	:	Embolemidae
	:	Plumaridae
	:	Sclerogibbidae
	:	Scolebythidae
Super Family	:	Vespoidea
Families	:	Bradynobaenidae
	:	Formicidae
	:	Mutillidae
	:	Pompilidae
	:	Rhopalsomatidae
	:	Sapygidae
	:	Scolidae
	:	Sierolomorphidae
	:	Tiphidae
	:	Vespidae

Infraorder: Parasitica

Super Family	:	Ceraphronoidea
Families	:	Ceraphronidae
	:	Megaspilidae
Super Family	:	Chalcidoidea
Families	:	Agaonidae
	:	Aphelinidae
	:	Chalcididae
	:	Encyrtidae
	:	Eucharitidae
	:	Eulophidae
	:	Eupelmidae
	:	Eurytomidae
	:	Leuscospidae
	:	Mymaridae
	:	Ormyridae
	:	Perilampidae
	:	Pteromalidae
	:	Rotoitidae
	:	Signiphoridae
	:	Tanaostigmatidae
	:	Tetracampidae
	:	Torymidae
	:	Trichogrammatidae
Super Family	:	Cynipoidea
Families	:	Austrocynipidae
	:	Cynipidae
	:	Figitidae
	:	Ibaliidae
	:	Liopteridae
Super Family	:	Evanioidea
Families	:	Aulacidae
	:	Evaniidae
	:	Gasteruptiidae

Super Family	:	Ichneumonoidea
Families	:	Braconidae
	:	Ichneumonidae
Super Family	:	Megalyroidae
Family	:	Megalyridae
Super Family	:	Mymarommatoidea
Family	:	Mymarommatidae
Super Families	:	Platygastroidea
	:	Platygastridae
	:	Scelionidae
Super Family	:	Proctotrupoidea
Families	:	Austroniidae
	:	Diapriidae
	:	Heloridae
	:	Maomingidae
	:	Monomachidae
	:	Pelecinidae
	:	Peradeniidae
	:	Proctotrupidae
	:	Roproniidae
	:	Vanhorniidae
Super Family	:	Stephanoidea
Family	:	Stephanidae
Super Family	:	Trigonaloidea
Family	:	Trigonalidae.

Family Formicidae is characterized by:

1. Wingless worker caste.
2. Head prognathus.
3. Antennae usually composed of 4-12 segments in female caste and 9-13 segments in males.
4. Antennae geniculate between the long basal segments (scape) and the remaining funicular segments and the first segment are often long.
5. Second abdominal segment reduced, forming a node (petiole) which is isolated from the alitrunk in front and the remaining abdominal segments behind.

6. The third abdominal segment also reduced and isolated (post petiole).
7. Metapleural gland present on alitrunk, opening above the metacoxa.

The family Formicidae contains 9000 species reported throughout the world. The family Formicidae is subdivided into 14 sub-families *viz.*, Aneuretinae, Apomyrminae, Cerapachyinae, Dolichoderinae, Dorylinae, Ecitoninae, Formicinae, Leptanillinae, Leptanilloidinae, Myrmeciinae, Myrmicinae, Nothomyrmecinae, Ponerinae, Pseudomyrmecinae. Out of which, following 5 were selected for the present work since they are more economic important.

1. Formicinae
2. Ponerinae
3. Dolichoderinae
4. Pseudomyrmecinae
5. Myrmicinae

Characteristics of above five sub-families are given.

The sub-family Dolichoderinae shows following characters:

1. Body with single reduced or isolated segment (petiole).
2. Apex of gaster with a semicircular to circular acidopore formed from hypopygium.
3. Sting vestigial or absent.
4. Hypopygium with its lateral margins smooth and without spines.
5. Tergite of helcium with a U or V- shaped emargination dorsally in its anterior margin.

The sub-family Formicinae shows following characters:

1. Body with single reduced or isolated segment (petiole) between alitrunk and gaster.
2. Constriction between two basal abdominal segments.
3. Apex of gaster with circular acidopore formed from hypopygium.
4. Acidopore projects as a nozzle, which is fringed with setae.
5. Clypeus broad.
6. Frontal carinae present.
7. Eyes usually present only rarely absent.
8. Antennae with 8-12 segments.
9. Metapleural gland present and absent in some species.
10. Sting absent.

The sub-family Myrmecinae shows following characters:

1. Pedicel distinctly two- jointed in all sexes.
2. Presence of eyes.

3. Antennae long, thick close to the anterior margin of the head.
4. Well separated antennal socket.
5. Sting is present but not exerted.
6. Transversely rounded and unarmed pygidium.

The sub-family Ponerinae shows following characters:

1. Constriction between the basal 2 segments of the abdomen.
2. Unmodified powerful sting.
3. The body is more or less elongate and cylindrical, the abdomen especially.
4. The mandibles powerful.
5. Antennae more or less massive.
6. Eyes present, absent in one or two genera.
7. Pedicel one jointed.
8. Legs moderately long.
9. Most of the members are black.

The sub-family Pseudomyrmecinae shows following characters:

1. Body slender.
2. Eyes large.
3. Short mandibles.
4. Flexible promesonotal connection.
5. Presence of a post petiole.
6. Antennal sockets partly exposed in full-face (frontal) view.
7. The scape is relatively short.
8. Clypeus is narrow (front to back) and does not extend posteriorly between the frontal carinae.
9. The metapleural gland orifice is situated at the extreme posteroventral margin of the metapleuron.
10. The hind tibia usually has two apical spurs, of which the posterior spur is pectinate; and the sting is well developed.

Sub-family Formicinae

It contains 11 tribes and 51 genera.

Tribe Lasiini

It contains 9 genera including ***Paratrechina* Motschoulsky.**

Genus – *Paratrechina* Motschoulsky, 1863

According to Bolton (2012) 4 species of this genus have been reported from the world and 5 species were reported from India.

The genus is characterized by,

1. Antenna with 9-12 segments.
2. Mandibles sub triangular not extending into long slender blades.
3. Antennal sockets situated close to the posterior clypeal margin.
4. Metapleuron with metapleural gland situated above the hind coxa.
5. The head in full- face view the eyes at or infront of the midlength of the sides.
6. Head and alitrunk with setae arranged in pairs.

Key to Indian Species of the Genus *Paratrechina*

1. Mandibles with 8 teeth..*kohli*
   Mandibles with 5 teeth..2
2. Scape without macrosetae, dorsum of propodeum flat....................... *longicornis*
   Scape with macrosetae, dorsum of propodeum rounded 3
3. Macrosetae dark brown ....*zanjensis*
   Macrosetae yellow to whitish 4
4. Dark brown, cuticle giving greenish- blue reflection...............................*ankarana*
   Brown with lighter cuticle, propodeum smooth and shiny .... *antsingy*

Paratrechina ankarana Lapolla and Fisher, 2014

(Plate9, Figures 27 to 31)

Worker (Figure 27 and 28)

Body 2.9 mm long; cuticle giving greenish, blue reflection, head 0.6 mm long, 0.5 mm broad; antenna 12 segmented; thorax 1.1 mm long; hind leg 12.8 mm long; abdomen 0.9 mm long, 0.8 mm broad.

Head (Figure 29)

0.6 mm long, 0.5 mm broad, dark brown, cephalic index 83.33, oval, posterolateral corners; mandibles sub triangular to triangular, 0.1 mm long, with 5 teeth, mandibular index 20; eyes black,0.2 mm long, convex, infront of the midlength of the sides of head, with setae arranged with pairs; 3 small ocelli present.

Antenna (Figure 30)

4.95 mm long, 12 segmented including scape, scape 1.15 mm long, 0.1 mm broad, yellowish, scape index 191; pedicel 0.3 mm long, 0.1 mm broad, yellowish; flagellum 3.5 mm long, 10 segmented, antennal sockets situated close to the posterior clypeal margin, apical segments of antennal funiculus not forming a club.

Flagellar Formula

1 L/W = 4, 3 L/W = 3, 4 L/W = 4, A = 3.6

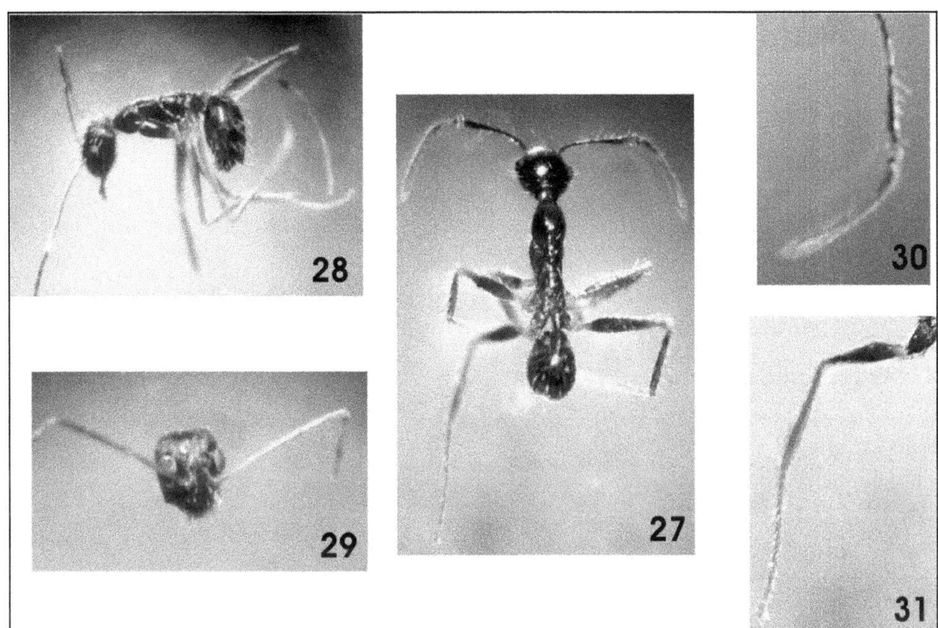

Plate 9: *Paratrechina ankarana*. Figure 27: Dorsal view; Figure 28: Lateral view; Figure 29: Head; Figure 30: Antenna; Figure 31: Hind leg.

Thorax

1.1 mm long, dark brown; in lateral view pronotum rises in a straight towards mesonotum; metapleuron with a distinct metapleural gland situated above hind coxa.

Wings – Absent.

Fore Leg

7.3 mm long, brown, coxa 0.4 mm long,brown; trochanter 0.1 mm long, yellow; femur 2.3 mm long, brown; tibia 1.9 mm long with spur; tarsus 1.4 mm long, brown; pre-tarsus 1.2 mm long (4- segments).

Mid Leg

7.1 mm long, brown, coxa 0.5 mm long, brown; trochanter 0.2 mm long, yellow; femur 2 mm long, brown; tibia 1.6 mm long; tarsus 1.6 mm long, brown; pre-tarsus 1.2 mm long (4- segments).

Hind Leg (Figure 31)

8 mm long, brown, coxa 0.5 mm long, brown; trochanter 0.2 mm long, yellow; femur 2 mm long, brown; tibia 2.2 mm long with spur; tarsus 1.9 mm long, brown; pre-tarsus 1.2 mm long (4- segments).

Abdomen

0.9 mm long, 0.8 mm broad, dark brown; orifice present, circular to sub circular

Sting – Absent.

Colour

Dark brown - Head, eyes, thorax, legs, abdomen.

Yellow- Antenna, trochanter

Host – Unknown

Host plant - Unknown

Paratype – 10 workers, Coll. Kurane, S.H. July. 2012 to Dec.2014, head, antenna, leg and abdomen mounted on card sheet and labeled as above.

Distributional Record

3 workers Karad, 1-VII-2012; 2 Shirala, 4-VIII-2012; 3Walwa, 11-X-2013; 2 Ajara, 11-VIII-2014

Tribe Plagiolepidini

It contains 7 genera including *Anoplolepis* Santschi.

Genus – *Anoplolepis* Santschi, 1914

The genus *Anoplolepis* is a small genus. Only nine species of this genus have been reported from the world (Bolton, 2012) and 1 species reported from Oriental region.

1. Antenna with 9-12 segments.
2. Antennal club distinct.
3. Eyes present.
4. Propodeum and petiole unarmed, without spines.
5. Mesonotum fused with metanotum, not separated by transverse groove.
6. Sting absent.

Key to Species of the Genus *Anoplolepis*

It has been given by Bolton (2012).

Anoplolepis gracilipes Smith, 1857

(Plate 10, Figures 32 to 36)

Worker (Figures 32 and 33)

Body 3.1 mm long; head 0.2 mm long, 0.3 mm broad; antenna 11 segmented; thorax 1.2 mm long; hind leg 12.9 mm long; abdomen 1.7 mm long, 1 mm broad.

Head (Figure 34)

0.2 mm long, 0.3 mm broad, orange red, cephalic index 150; mandibles triangular, 0.4 mm long, mandibular index 200; eyes 0.4 mm large, break outline of head, black; antennal sockets and posterior clypeal margin separated by a distance equal to or greater than minimum length of antennal scape.

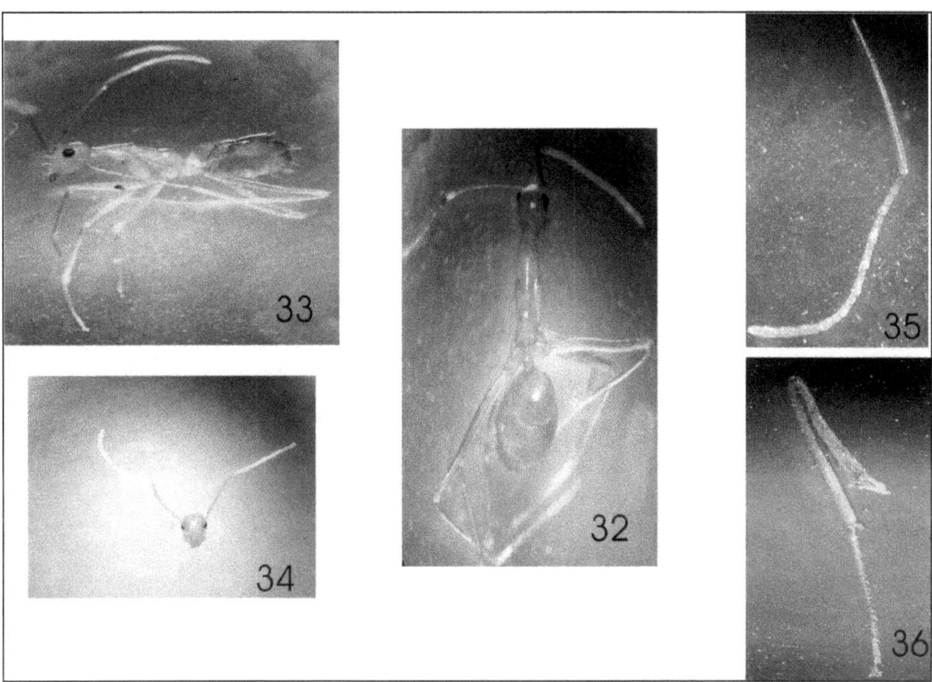

Plate 10: *Anoplolepis gracilipes*. Figure 32: Dorsal view; Figure 33: Lateral view; Figure 34: Head; Figure 35: Antenna; Figure 36: Hind leg.

Antenna (Figure 35)

8.3 mm long, yellow,11 segmented including scape, scape 4.9 mm long, 0.2 mm broad, yellow, scape length greater than head length scape index 1633.3; pedicel 0.3 mm long, 0.1 mm broad; flagellum 3.1 mm long, 9 segmented.

Flagellar Formula

1 L/W =2, 3 L/W = 3.5, 4 L/W = 3.5, A= 3.

Thorax

1.2 mm long, yellow, long, pronotum fused with metanotum, not separated by transverse groove; metapleuron with distinct gland orifice.

Fore Leg

11.5 mm long, pale yellow; coxa 1.5 mm long, pale yellow; trochanter 0.4 mm long, pale yellow; femur 3.6 mm long, pale yellow; tibia 3.1 mm long, pale yellow, with spur; tarsus 1.5 mm long, pale yellow; pre-tarsus 1.4 mm long (4 – segmented), pale yellow coloured.

Mid Leg

11 mm long, pale yellow; coxa 0.7 mm long, pale yellow; trochanter 0.4 mm long, pale yellow; femur 3 mm long, pale yellow; tibia 3.4 mm long, pale yellow;

tarsus 2 mm long, pale yellow; pre-tarsus 1.5 mm long (4 – segmented), pale yellow coloured.

Hind Leg (Figure 36)

12.9 mm long, pale yellow; coxa 0.8 mm long, pale yellow; trochanter 0.3 mm long, pale yellow; femur 3.8 mm long, pale yellow; tibia 3.8 mm long, pale yellow; tarsus 2.5 mm long, pale yellow; pre-tarsus 1.7 mm long (4 – segmented), pale yellow coloured.

Abdomen

Abdomen 1.7 mm long, 1 mm broad, 6- segmented, yellowish brown; petiole (abdominal segment 2), 0.5 mm long, upright and not appearing flattened; propodeum and petiole node lacking a pair of short teeth; gaster armed with acidopore, hairs not long, thick.

Sting – Absent.

Colour

Yellowish brown: Head, abdomen.

Pale yellow: Antenna, thorax, legs.

Black: Eyes.

Host plant: *Peltophorum pterocarpum* Heyne.

Paratype: 5 workers, Coll. Kurane, S.H. Jan. 2012 to Dec.2014, head, antenna, leg, abdomen mounted on the card sheet and labeled as above.

Distributional Record

2 workers Shahuwadi, 3-I-2012; 3 Kolhapur, 6-X-2012; 1 Sangli, 25-XII-2014

Tribe Oecophyllini

It contains 1 genus, ***Oecophylla* Smith.**

Genus – *Oecophylla* Smith, 1860

This genus is a very common and abundant in species distribution. The species of this genus found in Sub-Saharan Africa, southern India, south east Asia, and Australia. According to Bolton (2012) 2 species have been reported from world and 1 species from Oriental region.

1. An elongated first funicular segment.
2. Presence of propodeal lobes.
3. Helcium at midlength of abdominal segment 3.
4. Gaster capable of reflexion over the mesosoma.
5. Head quadrangular.
6. Mandibles long.
7. Clypeus convex.
8. Antennal carina short.

9. Antenna 12-jointed.
10. Eyes large.
11. Thorax elongate, pronotum convex
12. Legs long and slender.
13. Abdomen short, oval.

***Oecophylla smaragdina* Fabricius, 1775**

(Plate11, Figures 37 to 41)

Worker (Figure 37 and 38)

9 mm long; head 2 mm long, 1 mm broad; thorax 3 mm long; hind leg 6.6 mm long; abdomen 4 mm long 2 mm broad.

Head (Figure 39)

2 mm long, 1 mm broad, pale yellow; cephalic index 50; longer than broad; mandibles 0.5 mm long, triangular, mandibular index 25; clypeus broad; frontal carinae rising from posterior border of clypeus; eyes 1 mm long, present less than one half the side of the head back.

Antenna (Figure 40)

18 mm long, pale yellow;12 segmented including scape, scape 8 mm long, 0.2 mm broad, pale yellow; scape index 160; pedicel 2.5 mm long, 0.2 mm broad; flagellum 7.5 mm long,10 segmented.

Flagellar Formula

1 L/W = 3.3, 3 L/W = 3.3, 4 L/W = 2.6, A = 3.

Thorax

3 mm long, pale yellow, pro- meta and mesonotal sutures distinct, meta pleuron lacking metapleural gland absent.

Wings – Absent.

Fore Leg

18.4 mm long, pale yellow; coxa 1.6 mm long, pale yellow; trochanter 0.5 mm long, pale yellow; femur 6.4 mm long, pale yellow; tibia 5.4 mm long, pale yellow, with spur; tarsus 2.5 mm long, pale yellow; pre-tarsus 2 mm long (4 – segmented, pale yellow coloured.

Mid Leg

19.4 mm long, pale yellow; coxa 1.5 mm long, pale yellow; trochanter 0.5 mm long; femur 5.9 mm long, pale yellow; tibia 6 mm long, pale yellow, with spur; tarsus 3 mm long, pale yellow; pre-tarsus 2.5 mm long (4- segmented), pale yellow coloured.

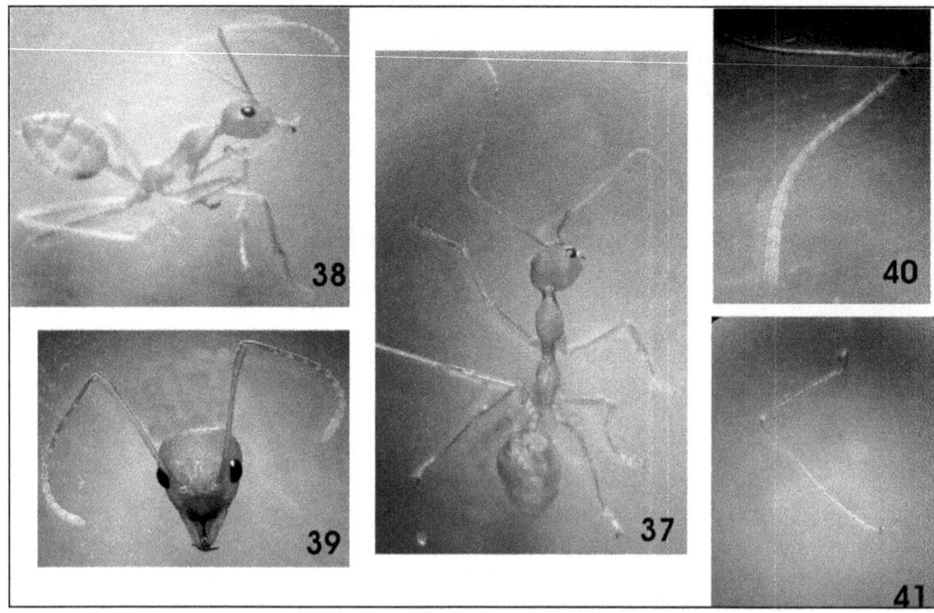

Plate 11: *Oecophylla smaragdina*. Figure 37: Dorsal view; Figure 38: Lateral view; Figure 39: Head; Figure 40: Antenna; Figure 41: Hind leg.

Hind Leg (Figure 41)

20.5 mm long, pale yellow; coxa 2.0 mm long, pale yellow; trochanter 0.2 mm long, pale yellow, short; femur 1.6 mm long, pale yellow; tibia 9.7 mm long, with spur, pale yellow; tarsus 4 mm long, pale yellow; pre-tarsus 3 mm long (4 – segmented), pale yellow coloured.

Abdomen

Abdomen 4 mm long, 2 mm broad, pale yellow, 6 segmented, oval, petiole; abdominal segment – 2, 1 mm long, pale yellow, reduced to an elongate low node; gaster reflexed over the alitrunk.

Sting – Absent.

Colour

Pale yellow: Head, thorax, antenna, leg, abdomen.

Eyes: Black.

Host- Many Lepidopterous Catterpillars on crop plants.

Host plant: *Saraca asoca* Roxb.

Paratype: 11 workers, Coll. Kurane, S.H. Feb. 2012 to Dec.2014, head, antenna, abdomen mounted on card sheet and labeled as above.

Distributional Record

5 workers Kolhapur, 7-II-2012; 3 Sangli, 25-II-2012; 3 Kagal 15-IV-2014

Tribe Camponotini

It contains 9 genera including *Camponotus* Mayr.

Genus – *Camponotus* Mayr, 1861

Camponotus is the largest genus in the world. The genus includes over 1,000 species (Bolton, 2012) and 47 have been reported from India. The genus shows following characters:

1. Antenna with 9-12 segments.
2. Mandibles sub triangular to elongate triangular.
3. Mandibles with 5-7 teeth.
4. Petiole an errect node.
5. Gaster not capable of reflexion over the alitrunk.
6. Eyes large, situated midlength of the sides.
7. Pronotum and petiole are armed with spines.
8. Apical segments of antennal funiculus not forming a club.
9. Metapleural gland orifice present above the hind coxa.
10. Abdomen oval.
11. Tergite of first gastral segment much smaller.

Key to Species of the Genus *Camponotus* (Bingham, 1903)

A. Thorax viewed from side forming a regular arch.

a. Pubescence on sides of head and long .... *barbatus*

b. Pubescence on sides of head and short, not forming a beard.

a1. Head, thorax and abdomen black

a2. Tibiae of the legs prismatic.

a3. Tibiae without spines .... *lamarcki*

b3. Tibiae of the legs with spines beneath.

a4. Abdomen covered with long yellowish hair .... *japonicus*

b4. Abdomen covered with sparse erect hairs.

a5. Maj. length 11-16 mm. Min. with head posteriorly narrow, not constricted to forma collar .... *compressus*

b5. Maj. length 17-21 mm. Min. with head posteriorly constricted to form a collar .... *angusticollis*

b2. Tibiae of the legs compressed but not prismatic.

a3. Abdomen with a fine thin pubescence.

a4. Length maj. below 7 mm; pubescence grey .... *binghamii*

b1. Length maj. over 9 mm; pubescence yellowish .... *paria*

b3. Abdomen opaque, without any fine sericeous pubescence.

a4. Hind tibiae spinous beneath...*dolendus*

b4. Hind tibiae without spines beneath .... *crassisquamis*

b1. Head, thorax and abdomen pale yellow... *invidus*

c1. Head, thorax and abdomen never all black or all yellow.

a2. Scape of antennae flat.

a3. Basal joint of tarsi broad and flat ....*mistura*

b3. Basal joint of tarsi narrow, slightly depressed.............................*fornaronis*

b2. Scape of antennae cylindrical.

a3. Abdomen with dense silk pubescence.

a4. Clypeus with a distinct medial lobe produced anteriorly ...... *rufoglaucus*

b4. Clypeus without a medial lobe, anterior margin transverse......... *mendar*

b3. Abdomen without, or with very thin, sparse recumbent pubescence.

a4. Tibiae cylindrical.

a5. Tibiae covered with long erect hairs........................*buddhae*

b5. Tibiae covered with very widely spaced hairs

a6. With a few spines on apical third of tibiae....................................*oblongus*

b6. Without spines of tibiae.

a7. Medial lobe of clypeus with its anterior margin rounded ....*wroughtoni*

b7. Medial lobe of clypeus with its anterior margin transverse.

a8. Maj. below 8mm, min. under 6 mm ....*taylori*

b8. Maj. over 8 mm, min. over 6 mm

a9. Maj. with 7 teeth, min with 6 teeth..*infuscus*

b9. Maj. With 6 teeth, min with 5 teeth ....*variegatus*

b4. Tibiae compressed.

a5. Tibiae spined

a6. Head, thorax and abdomen castaneous red.

a7. Maj.over 15mm, Min. over 10 mm..*festinus*

b7. Maj. not over 8, min. 5- 6 mm... *arrogans*

b6. Head and abdomen black or dark castaneous red; thorax yellow

a7. Medial lobe of clypeus long rectangular

a8. Head much broader posteriorly than infront ....*dichrous*

b8. Head only as a broad posteriorly as infront......................................*basalis*

b7. Medial lobe of clypeus short, the lateral angle rounded................ *irritans*

b5. Tibiae without spines

a6. Head, thorax and abdomen rugulose opaque.

a7. Castaneous brown.. *badius*

b7. Reddish yellow apex of abdomen darker................................ *nicobarensis*

b6. Head, thorax and abdomen sparsely punctured, shining, not opaque.

a7. Head, thorax and abdomen dark castaneous brown.

a8. Pronotum longer than mesonotum, strongly constricted, anteriorly forming a neck.. *carin*

b8. Pronotum about equal in length to mesonotum, slightly constricted, in front not forming a distinct neck .... *thoraso*

b7. Head and abdomen fuscous brown or black (in min. head sometimes yellow); thorax honey yellow.

a8. Distance between the antennal carinae equal to distance between eyes and antennal carina.. *milis*

b8. Distance between antennal carinae distinctly greater than between than eyes and antennal carinae .... *pallidus*

B. Regular arch of the thorax interrupted by the apex of the metanotum being truncate.

a. Mandibles toothed at apex and also on inner margin............................. *gigas*

b. Mandibles toothed only at apex.

a1. Clypeus anteriorly emarginate in the middle........................... *marginatus*

b1. Clypeus not emarginate.

a2. Scape of antennae flattened .... *radiatus*

b2. Scape of antennae cylindrical.

a3. Length of maj. over 12 mm, min. over 5mm .... *siemsseni*

b3. Length of maj. below 7mm, min. under 5 mm.

a4. Head, thorax and abdomen reddish brown............................... *reticulatus*

b4. Head and thorax black, abdomen castaneous ....*yerburyi*

C. Regular arch of the thorax interrupted by the metanotum being raised, rounded above and gibbous.

a. Anterior lateral angles of pronotum dentate or sub dentate.

a1. Abdomen with dense, recumbent, sericeous golden pile hiding the sculpture..*auriventris*

b1. Abdomen without recumbent pile,the sculpture plainly visible..*wasmanni*

b. Anterior, lateral angles of pronotum rounded, not dentate.

a1. Length over 9 mm.

a2. Thorax posteriorly and node of pedicel coarsely punctured .... *holosericeus*

b2. Thorax posteriorly and node of pedicel finely reticulate-punctate, rugulose .... *camelinus*

b1. Length under 9 mm .... *confucii*

D. Regular arch of the thorax interrupted at the meso-metanotal suture by the metanotum forming an angle with the mesonotum; basal portion of metanotum horizontal, flat or slightly concave; apical portion excavate.

a. Tibiae of legs spinous beneath.

a1. Length 6-10 mm;node of pedicel thick .... *sericeus*

b1. Length 3-4mm; node of pedicel broader than long flat .... *varians*

b. Tibiae of legs not spinous beneath .... *nirvanae*

Camponotus compressus Fabricius, 1787

(Plate 12, Figures 42 to 46)

Worker (Figure 42 and 43)

Body 8 mm long, head 2 mm long, 2.1 mm broad; antenna 12 segmented; thorax 2.5 mm long; hind leg 20.6 mm long, moderately long; abdomen 3.5 mm long, 1 mm broad.

Head (Figure 44)

2 mm long, 2.1 mm broad, brown, posteriorly narrow; cephalic index 91.66; eyes present, black, 1 mm long; clypeus broad; frontal carina present; mandibles triangular, with 5 teeth, mandibular index 11.36.Head covered with short erect hairs.

Antenna (Figure 45)

13.5 mm long,pale yellow, 12 segmented, hairy; scape 6 mm long,0.3 mm broad, pale yellow; pedicel 1 mm long, 0.3 mm broad, pale yellow; flagellum 6.5 mm long,10 segmented, apical segments of antennal funiculus not forming a club.

Flagellar Formula

1 L/W = 2.6, 3 L/W = 3, 4 L/W = 3, A = 2.8

Thorax

2.5 mm long, brownish yellow, covered with erect hairs, constriction at the meso – metanotal suture; metapleural gland absent;

Wings – Absent.

Fore Leg

17.3 mm long, pale yellow,covered with erect hairs; coxa 3 mm long, pale yellow; trochanter 0.5 mm long,pale yellow; femur 4.9 mm long, pale yellow; tibia

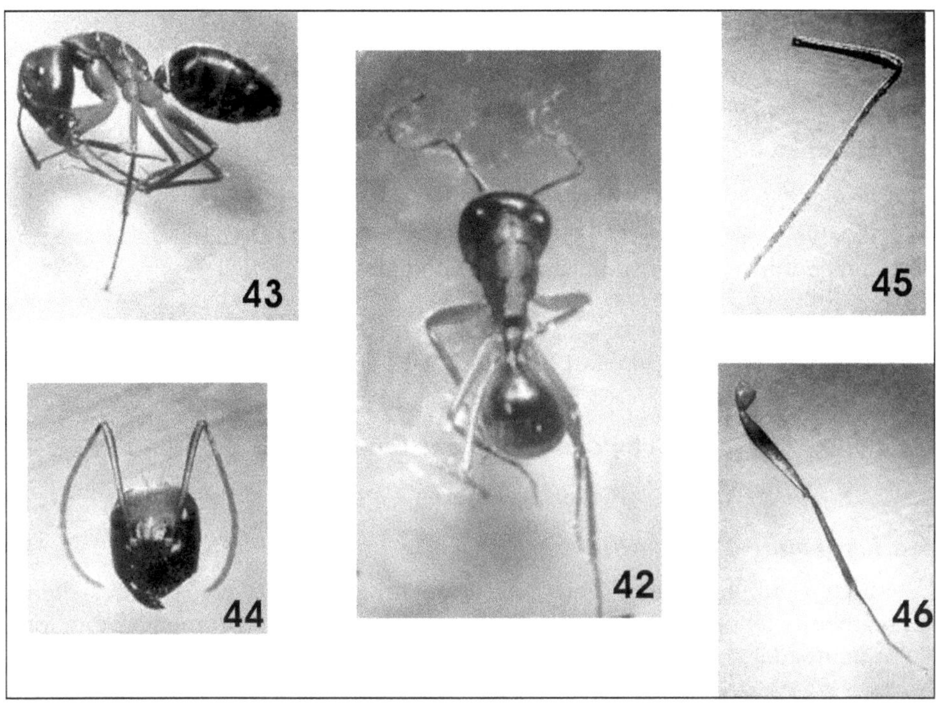

Plate 12: *Camponotus compressus*. Figure 42: Dorsal view; Figure 43: Lateral view; Figure 44: Head; Figure 45: Antenna; Figure 46: Hind leg.

3.9 mm long, tibia with spur; tarsus 3 mm long,pale yellow; pre-tarsus 2 mm long (4-segmented).

Mid Leg

16 mm long, pale yellow,covered with erect hairs; coxa 1.5 mm long, pale yellow; trochanter 0.5 mm long pale yellow; femur 5 mm long, pale yellow; tibia 4 mm long, pale yellow;tarsus 3 mm long pale yellow; pre-tarsus 2 mm long (4 – segmented).

Hind Leg (Figure 46)

20.6 mm long, pale yellow; covered with erect hairs; coxa 1.5 mm long, pale yellow; trochanter 0.6 mm long; femur 6.5 mm long, pale yellow; tibia 7 mm long, tibia with spur; tarsus 3 mm long, pale yellow; pre-tarsus 2 mm long (4 – segmented).

Abdomen

Abdomen 3.5 mm long, 1 mm broad, testaceous, shining, 6 segmented, short, globose; petiole (abdominal segment – 2), 1 mm long, brown, abdominal segments with sclerites.

Sting – Absent.

Colour

Black: Head, abdomen.

Testaceous brown: Antenna, thorax, legs.

Host plant : *Tectona grandis* L.

Hosts – Cicadellids, Scales, Delphacids.

Paratype –10 workers, Coll. Kurane, S.H. April. 2012 to Dec.2014, head, antenna, leg, and abdomen mounted on card sheet and labeled as above.

Distributional Record

3 workers Shirala, 25-IV-2012; 2 Palus, 21-VI-2012; 3 Sangli, 24-II-2013; 2 Karad, 29-V-2013.

Camponotus irritans Smith, 1858

(Plate 13, Figures 47 to 51)

Worker (Figure 47 and 48)

Body 7 mm long; head 2 mm long, 1 mm broad; pilosity pale yellow; antenna 12 segmented; thorax 2.5 mm long; hind leg 21.1 mm long; abdomen 2.5 mm long, 1.1 mm broad.

Head (Figure 49)

Castaneous brown, 2 mm long, 1 mm broad; cephalic index 50; eyes 1 mm long, large, black; mandibles 0.7 mm long, mandibular index 35; shining,dark, sub triangular; mandibles sub triangular; clypeus black; medial lobe short lateral angle rounded.

Antenna (Figure 50)

12.4 mm long, brownish yellow, 12 segmented including scape; scape 4.8 mm long, 0.3 mm broad; scape index 480; pedicel 0.6 mm long, 0.3 mm broad, brownish yellow; flagellum 7 mm long, 10 segmented, apical segments not forming a club.

Flagellar Formula

1 L/W = 3, 3 L/W = 3.3, 4 L/W = 3.3, A = 3.2

Thorax

2.5 mm long, yellow, longer, narrower; metapleuron lacking metapleural gland.

Fore Leg

17.2 mm long, yellowish; coxa 2.9 mm long, yellowish; trochanter 0.5 mm long, yellowish; femur 4.7 mm long, yellowish; tibia 3.7 mm long, with spur; yellowish; tarsus 3.5 mm long, yellowish; pre –tarsus 1.9 mm long (4 – segmented).

Mid Leg

13.8 mm long, yellowish; coxa 1 mm long, yellowish; trochanter 0.5 mm long, yellowish; femur 4.4 mm long, yellowish; tibia 4 mm long, yellowish; tarsus 1.9 mm

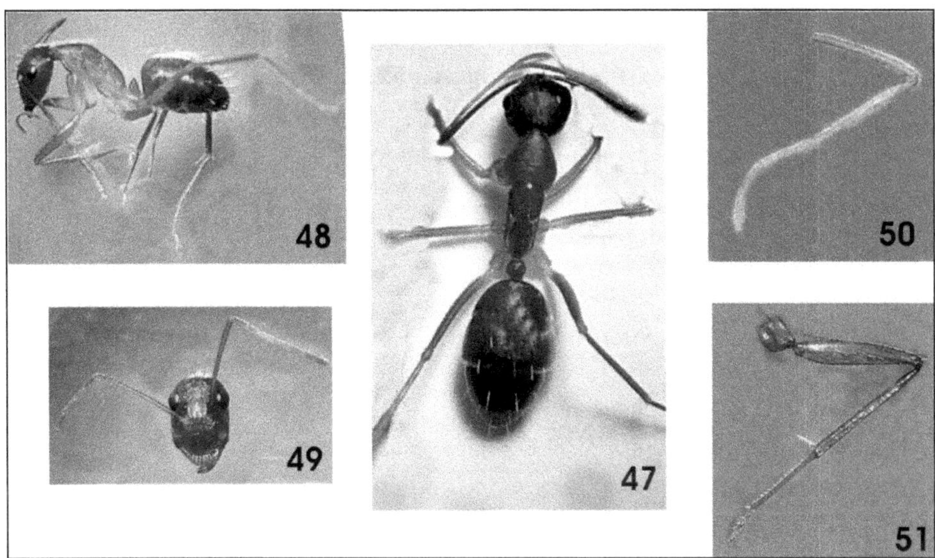

Plate 13: *Camponotus irritans*. Figure 47: Dorsal view; Figure 48: Lateral view; Figure 49: Head; Figure 50: Antenna; Figure 51: Hind leg.

long, yellowish; pre- tarsus 2 mm long, yellowish (4 – segmented).

Hind Leg (Figure 51)

21.1 mm long, yellowish; coxa 1.9 mm long, yellowish; trochanter 0.7 mm long, yellowish; femur 6 mm long, yellowish; tibia 6.3 mm long, yellowish, with spur; tarsus 3.5 mm long, yellowish; pre-tarsus 2.7 mm long, yellowish (4 – segmented).

Abdomen

Abdomen 2 mm long, 1.1 mm broad; petiole (abdominal segment 2), 0.5 mm long, nodiform, thick scale like; tergite smaller.

Sting – Absent.

Colour

Castaneous brown – Head, antenna, thorax.

Yellowish – Legs.

Castaneous brown – Abdomen.

Host plant – *Zizyphus jujuba* Miller, 1768.

Host- Scales insects, some Rodents.

Paratype –9 workers, Coll. Kurane, S.H. Jan. 2012 to Dec.2014, head, antenna, leg, and abdomen mounted on card sheet and labeled as above.

Distributional Record

2 workers Shahuwadi, 3-I-2012; 3 Shirala, 10-IV-2012; 2 Kagal, 25-X-2013; 2 Miraj, 14-X-2014.

Camponotus variegatus **Smith, 1858**

(Plate 14, Figures 52 to 56)

Worker (Figure 52 and 53)

Body 7.1 mm long; head 1.6 mm long, 1 mm broad; antenna 12 segmented; thorax 2.7 mm long; hind leg 19.5 mm long; abdomen 2.8 mm long, 1 mm broad.

Head (Figure 54)

1.6 mm long, 1 mm broad, darker yellowish red, sub triangular, cephalic index 83.33; mandibles 1 mm long, sub triangular with 5 teeth, mandibular index 83.33; eyes 0.7 mm, large, prominent and situated behind mid length of sides; clypeus tectiform.

Antenna (Figure 55)

12 mm long, yellow,12 segmented including scape, scape 5 mm long, 0.2 mm broad, pale yellowish red, antennal sockets situated far behind the posterior clypeal margin, scape length greater than head length, scape index 500; pedicel 1 mm long, 0.2 mm broad; flagellum 6 mm long, 10 segmented, apical segment of anterior funiculus not forming a club.

Flagellar Formula

1 L/W =2, 3 L/W = 2.3, 4 L/W = 2.3, A= 2.2

Thorax

2.7 mm long, yellowish red, narrow rather compressed.

Fore Leg

13.7 mm long, pale yellowish red; coxa 1.5 mm long, pale yellowish red;

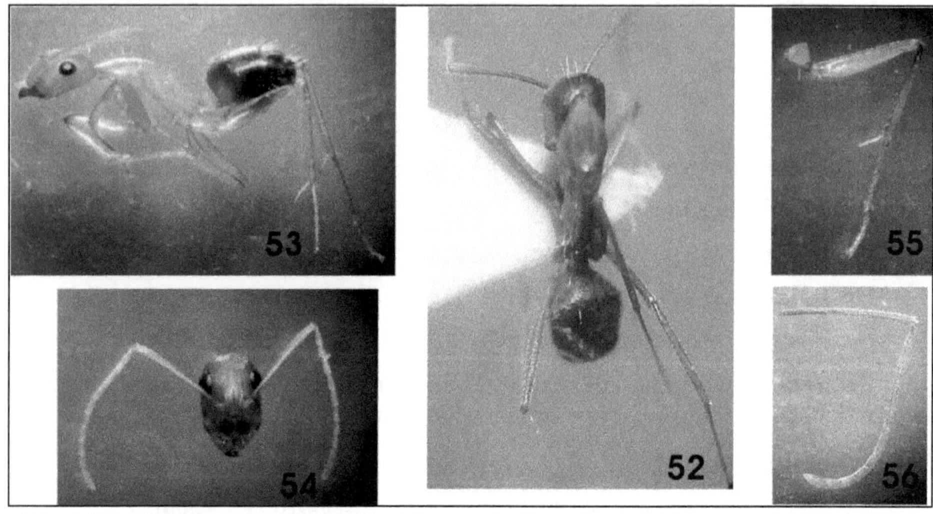

Plate 14: *Camponotus variegatus*. Figure 52: Dorsal view; Figure 53: Lateral view; Figure 54: Head; Figure 55: Antenna; Figure 56: Hind leg.

trochanter 0.3 mm long, pale yellowish red; femur 4.4 mm long, pale yellowish red; tibia 4.5 mm long, pale yellowish red, with spur; tarsus 1 mm long, pale yellowish red; pre-tarsus 2 mm long (4 – segmented), pale yellowish red.

Mid Leg

17.3 mm long, pale yellowish red; coxa 1.2 mm long, pale yellowish red; trochanter 0.4 mm long, pale yellowish red; femur 6.2 mm long, pale yellowish red; tibia 5.5 mm long, pale yellowish red; tarsus 1 mm long, pale yellowish red; pre-tarsus 3 mm long (4 – segmented), pale yellowish red.

Hind Leg (Figure 56)

19.5 mm long, pale yellowish red; coxa 1.5 mm long, pale yellowish red; trochanter 0.5 mm long, pale yellowish red; femur 6.4 mm long, pale yellowish red; tibia 6.5 mm long, pale yellowish red, with spur, cylindrical; tarsus 1.6 mm long, pale yellowish red; pre-tarsus 3 mm long (4 – segmented), pale yellowish red.

Abdomen

Abdomen 2.8 mm long, 1 mm broad, short, broadly oval, dark yellowish red, dark than thorax; the petiole (abdominal segment 2), 1.5 mm long, node of pedicel small.

Sting – Absent.

Colour

Dark yellowish red: Head, abdomen.

Pale yellowish red: Antenna, thorax, legs.

Black: Eyes.

Host plant: *Zizyphus jujuba* Miller, 1768.

Host- Scales insects, some Rhodents.

Paratype –6 workers, Coll. Kurane, S.H. Jan. 2012 to Dec.2014, head, antenna, leg, abdomen mounted on the card sheet and labeled as above.

Distributional Record

3 workers Shahuwadi, 3-I-2012; 2 Shirala, 10-IV-2012; 1 Patan, 7-X-2013.

***Camponotus sericeus* Fabricius, 1798**

(Plate 15, Figures 57 to 61)

Worker (Figure 57 and 58)

6.2 mm long; head 1.4 mm long, 1.5 mm broad; antenna 12 segmented; thorax 2.1 mm long; hind leg 20.7 mm long; abdomen 2.7 mm long, 1.1 mm broad.

Head (Figure 59)

1.4 mm long, 1.5 mm broad, black, cephalic index 107.14, with granular appearance, errect, pubescence, very broad, emarginated; clypeus broad, anterior border broadly emarginated in the middle; mandibles 1.8 mm long, triangular, mandibular index 128.7; eyes 1 mm long, black.

Antenna (Figure 60)

10.5 mm long, castaneous; antennal sockets situated well behind the posterior margin of the clypeus; 12 segmented including scape; scape 5 mm long, 0.3 mm broad, castaneous, scape index 33.33; pedicel 1 mm long, 0.3 mm broad, castaneous; flagellum 4.5 mm long, 10 segmented.

Flagellar Formula

1L/W = 2.3, 3 L/W = 2.3, 4 L/W = 2.3, A = 2.3.

Thorax

2.1 mm long, black, broad in front, compressed posteriorly, emarginated at the meso – metanotal suture, metanotum and its spiracles present on dorsal alitrunk; metapleural gland absent.

Wings – Absent.

Fore Leg

15.5 mm long, castaneous, stout; coxa 3.4 mm long, castaneous; trochanter 0.5 mm long, castaneous; femur 4 mm long,castaneous; tibia 3 mm long, cylindrical, castaneous, with spur; tarsus 2.3 mm long, castaneous; pre-tarsus 2.3 mm long, castaneuos (4 – segmented).

Plate 15: *Camponotus sericeus*. Figure 57: Dorsal view; Figure 58: Lateral view; Figure 59: Head; Figure 60: Antenna; Figure 61: Hind leg.

Mid Leg

14.1 mm long, castaneous, stout; coxa 1.5 mm long, castaneous; trochanter 0.5 mm long, castaneous; femur 2.7 mm long, castaneous; tibia 3.6 mm long, cylindrical, castaneous; tarsus 3.1 mm long, castaneous; pre-tarsus 2.7 mm long, castaneous (4 – segmented).

Hind Leg (Figure 61)

20.7 mm long, castaneous, stout; coxa 1.4 mm long, castaneous; trochanter 0.7 mm long, castaneous; femur 5 mm long, castaneous; tibia 6.4 mm long, cylindrical, castaneous, with spur; tarsus 4 mm long, castaneous; pre-tarsus 3.2 mm long, castaneous (4 – segmented).

Abdomen

Abdomen 2.7 mm long, 1.1 mm broad, with a dense recumbent silky golden pubescence, broad, globose; the petiole (abdominal segment 2), 1 mm long; pygidium large.

Acidopore – Present.

Colour

Black – Head, thorax.

Dark custaneous – Antenna, legs.

Silky golden – Abdomen.

Host - Unknown.

Paratype – 2 workers, Coll. Kurane, S.H. April. 2012 to Oct.2014, head, antenna, legs, abdomen mounted on card sheet and labeled as above.

Distributional Record

2 worker Shirala, 25-IV-2012.

Genus – *Polyrachis* Smith, 1857

This genus is found in the World with 697 species (Bolton, 2021). In India 48 species have been reported. The genus shows following characters,

1. There is little difference in the size and form of workers.
2. Apical segments of antennal funiculus not forming a club.
3. Spines present on pronotum, propodeum and petiole.
4. Abdomen short.
5. Metapleural gland absent.
6. Tergite of first gastral segment large.
7. Basal segment covering more than half of the total length of the abdomen.

Key to Species of the Genus *Polyrachis* (Bingham, 1903)

A. Thorax and pedicel armed with spines.

a. Thorax rounded above, the sides not margined along their whole length.

a1. Pro-and mesonotum with a spine on each side.

a2. Spines on node of pedicel parallel for a part of their length from base.

a3. Pronotal spines pointing outwards, curved laterally backwards, forming hooks .... *bihamatat*

b3. Pronotal spines pointing outwards, slightly bent downwards, not forming hooks .... *bellicosa*

b2. Spines on node of pedicel not parallel, divergent from base .... *ypsilon*

b1. Pro-and metanotum with a spine on each side, mesonotum unarmed.

a2. Pubescence soft, erect and abundant.

a3. Spines on node of pedicel forming hook .... *furcata*

b3. Spines on node of pedicel not forming hooks.

a4. Head not punctured, smooth, shining .... *gracilior*

b4. Head coarsely punctured posteriorly .... *phipsoni*

b2. Pubescence short,silky and recumbent, erect or entirely absent.

a3. Metanotal spines forming hooks.

a4. Abdomen with dense golden pile .... *rupicapra*

b4. Abdomen without pile or pubescence.

a5. Thorax finely punctured,head and abdomen opaque .... *hodgsoni*

b5. Thorax coarsely punctured, head and abdomen shining .... *arachne*

b3. Metanotal spines not forming hooks.

a4. Basal portion of metanotum not margined laterally.

a5. Pubescence sparse, entirely wanting.

a6. Pedicel spines wide spreading, shaped so as to encircle front of abdomen.

a7. Head with tubercle on each side behind the eyes .... *tubericeps*

b7. Head not tuberculate .... *thompsoni*

b6. Pedicel spines not so wide spreading,not shaped so as to encircle the abdomen.

a7. Head, thorax and abdomen shining metallic blue or purple .... *venus*

b7. Head, thorax and abdomen black and sometimes red.

a8. Head, thorax and node of pedicel coarsely punctured.

a9. Abdomen black; length 9-10.5 mm .... *armata*

b9. Abdomen ferruginous; length 5.7mm..*fortis*

b8. Head, thorax and node of pedicel finely punctured.

a9. Node of pedicel with 2 median vertical short spines between spines between spines on upper lateral angles of node ....*haunwelli*

b9. Node of pedicel without median spines.......................................*simplex*

b3. Pubescence dense, silky and recumbent.

a6. Abdomen red ....*bicolar*

b6. Abdomen black.

a7. Pubescence bronzy yellow or golden.

a8. Two small teeth between spines on upper lateral angles of node of pedicel.. *dives*

b8. Three small teeth between spines on upper lateral angles of node of pedicel ....*affinis*

b7. Pubescence silvery ....*tibialis*

b4. Basal portion of metanotum distinctly margined laterally.

a3. Pronotal and metanotal spines subequal.

a6. Length 7-9 mm.

a7. Abdomen steel-blue...*chalybea*

b7. Abdomen red ....*abdiminalis*

b2. Length 6-7 mm; abdomen bronze green.. *aedipus*

b5. Metanotal spines nearly twice the length of the pronotal spines.

a6. Mesonotum concave; tibiae with spines on the inner margin........ *mutata*

b6. Mesonotum convex; tibiae without spines on the inner margin .... *binghamii*

c1. Pronotum with a short tooth;metanotum with a spine on each side; mesonotum unarmed.

a2. Abdomen not depressed, convex above

a3. Length 6-7 mm; abdomen red ....*levigata*

b3. Length 4-5 mm; abdomen black.. *ceylonensis*

b2 Abdomen strongly depressed, slightly convex above.............*wroughtoni*

d1. Pronotum with a short tooth on each side; mesonotum and metanotum unarmed ....*laevissima*

b. Thorax more or less flat above, the sides margined along their whole length.

a1. Pronotum, mesonotum and metanotum with a spine on each side... *craddocki* b1. Pronotum with a spine, mesonotum and metanotum with a triangular lamina on each side .... *horni*

c1. Pronotum with a long spine; mesonotum unarmed metanotum with a tubercle on each side

a2. Node of Pedicel with a 2 long spine on upper angles and two short lateral spines on sides.

a3. The lateral spines truncate or bimucronate at apex

a1. Pubescence very dense .... *proxima*

b1. Pubescence sparse .... *intermedia*

b3. The lateral spines pointed, not truncate nor bi mucronate at apex

a4. Pubescence very dense .... *mayri*

b4. Pubescence absent or very sparse

a5. Antennal carinae long, divergent posteriorly .... *striata*

b5. Antennal carinae short not divergent posteriorly.

a6. Legs thickly covered with long erect hairs .... *hamulata*

b6. Legs hairless, smooth .... *yerburyi*

b2. Node of pedicel armed with four short sub equal spines.

a3. Length 9-10 mm .... *striatorugosa*

b3. Length 5-6mm .... *covvexa*

d1. Pronotum with a short spine; mesonotum un armed; metanotum with a lamina spine on each side

a2. Metanotum with backward pointing laminate spine on each side, with the apices curved inwards, shaped like a pair of callipers ...... *selene*

b2. Metanotum with a vertical spines

a3. Node of pedicel with a two long spines on upper lateral angles and two short obtuse teeth between them .... *jerdoni*

b3. Node of the pedicel quadridenatate orquadrispinous, the spines typically subequal .... *punctillata*

c3. Node of pedicel trispinous.

a4. Antennal carinae distinctly divergent posteriorly .... *thrinax*

b4. Antennal carinae not divergent posteriorly .... *franenfeldi*

e1. Pronotum and mesonotum unarmed; metanotum with a spine on each side

a2. Metanotal spines broad triangular, pointing backward.

a3. Head and thorax punctured, not striate .... *clypeata* b3. Head and thorax striate.. *rastrata*

b2. metanotal spines vertical, very small.. *halidyi*

B. Thorax wholly unarmed; pedicel with four subequal spines ....*rastellata*

***Polyrachis dives* Smith, 1857**

(Plate 16, Figures 62 to 66)

Worker (Figure 62 and 63)

Body 7 mm long; head 1.1 mm long, 1 mm broad; antenna 12 segmented; thorax, 2 mm long; hind leg 21.4 mm long; abdomen 3.9 mm long, 2 mm broad.

Head (Figure 64)

1.1 mm long, 1 mm broad, black, cephalic index 90.90, oval, narrowed posteriorly; mandibles triangular, 1 mm long, mandibular index 17.85; clypeus broad from front to back; frontal carinae present; eyes 1 mm long, large, situated behind midlength of sides, black.

Antenna (Figure 65)

12.5 mm long, blackish, 12 segmented including scape; scape 6 mm long, 0.4 mm broad, blackish, scape index 600; pedicel 1 mm long, 0.4 mm broad, blackish; on upper lateral angles of pedicel node two small teeth between spines; flagellum 5.5 mm long, 10 segmented, apical segments of antennal funiculus not forming a club.

Flagellar Formula

1 L/W = 2, 3 L/W = 1.75, 4 L/W = 1.75, A = 1.8

Thorax

2 mm long, black, higher than the head, pronotal and metanotal spines long, pronotal spines pointing forwards, metanotal and pedicel spines pointing backwards.

Wings – Absent

Fore Leg

15.3 mm long, black; coxa 2.2 mm long,black; trochanter 0.5 mm long, black; femur 4.3 mm long, black; tibia black 3.5 mm long with spur; tarsus 2.5 mm long, black; pre-tarsus 2.3 mm long (4- segments).

Mid Leg

15.7 mm long, black; coxa 1 mm long,black; trochanter 0.5 mm long, black; femur 6 mm long,black; tibia 4 mm long, black; tarsus 2 mm long, black; pre-tarsus 2.2 mm long (4 – segments).

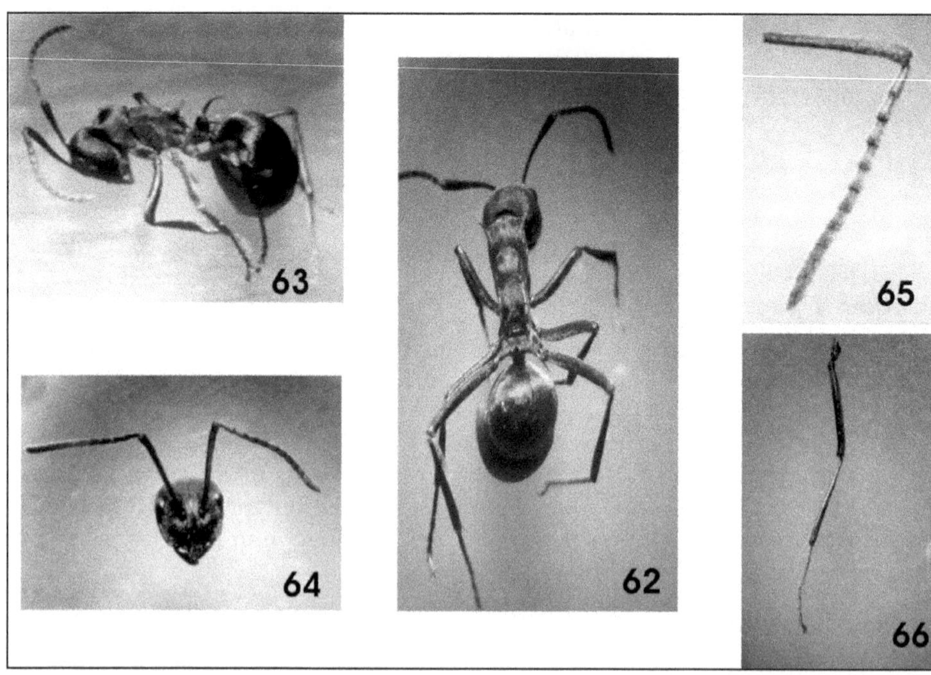

Plate 16: *Polyrachis dives*. Figure 62: Dorsal view; Figure 63: Lateral view; Figure 64: Head; Figure 65: Antenna; Figure 66: Hind leg.

Hind Leg (Figure 66)

21.4 mm long, black; coxa 1.4 mm long, black; trochanter 0.5 mm long, black; femur 6.5 mm long, black; tibia 6 mm long, black; tarsus 4 mm long, black; pre-tarsus 3 mm long (4 – segments).

Abdomen

Abdomen 3.9 mm long, 2 mm broad, 6 segmented,silky golden, globose; petiole (abdominal segment 2), with pair of spines, pointing backwards, 1.1 mm long; first gastral tergite large half the length of gaster, first gastral tergite much longer than the second.

Sting – Absent.

Colour

Black - Head, eyes, antenna, thorax, legs.

Golden - Abdomen.

Host plant – *Musa acuminata* Colla, 1820.

Host- Aphids.

Paratype – 4 workers, Coll. Kurane, S.H. July. 2012 to Dec.2014, head, antenna, leg and abdomen mounted on card sheet and labeled as above.

Distributional Record

1 worker Karad, 1-VII-2012; 2 Karad, 3-XI-2012; 1 Shirala, 4-IX-2013.

Polyrachis convexa Roger, 1863

(Plate 17, Figures 67 to 71)

Worker (Figure 67 and 68)

Body 4 mm long, covered with a short, recumbent glistening gray; head 0.8 mm long, 1 mm broad; antenna 12 segmented; thorax 2 mm long; hind leg 16 mm long; abdomen 1.2 mm long, 1 mm broad; lateral spines not truncate but pointed at apex.

Head (Figure 69)

0.8 mm long, 1 mm broad, black, short, as broad posteriorly as in front, cephalic index 12; mandibles sub triangular to elongate triangular not extended into long, slender blades, less than head length, 1 mm long, mandibular index 125; eyes 0.7 mm, large, prominent and situated behind mid length of sides; clypeus tectiform.

Antenna (Figure 70)

7 mm long, yellow,12 segmented including scape; scape 3.2 mm long, 0.4 mm broad, black, antennal sockets situated far behind the posterior clypeal margin, scape length greater than head length, scape index 320; pedicel 0.6 mm long, 0.3 mm broad; flagellum 3.2 mm long, 10 segmented, apical segment of anterior funiculus not forming a club.

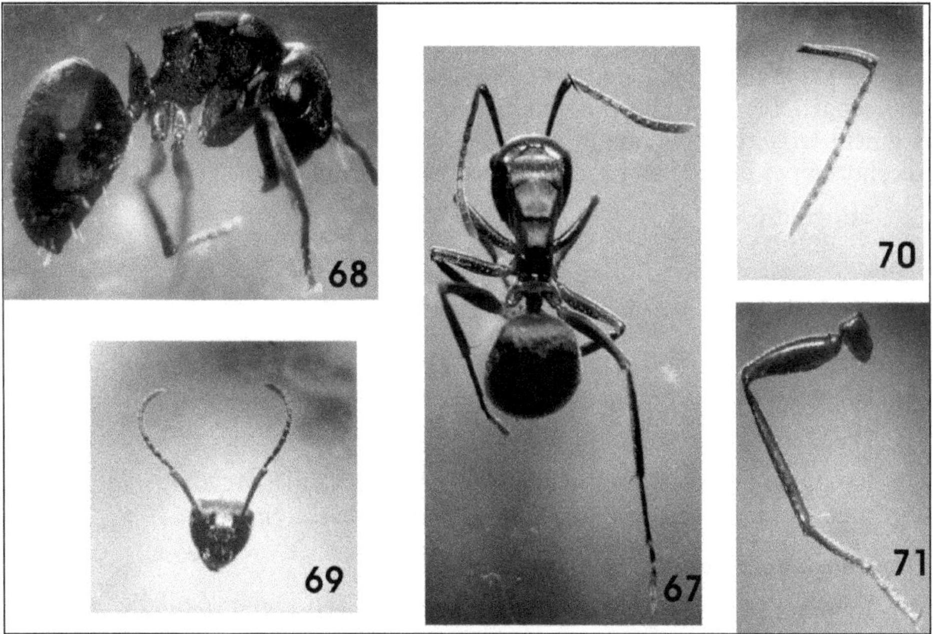

Plate 17: *Polyrachis convexa*. Figure 67: Dorsal view; Figure 68: Lateral view; Figure 69: Head; Figure 70: Antenna; Figure 71: Hind leg.

Flagellar Formula

1 L/W =1.5, 3 L/W = 1.5, 4 L/W = 1.5, A= 1.5

Thorax

2 mm long, black, strongly arched, the pro-, meso- and basal portion of metanotum deeply concave, pronotal spines short, acute, directed divergently forward, basal portion of metanotum bounded posteriorly by a slight carina between two short erect points at its lateral posterior angles.

Fore Leg

11.2 mm long, black; coxa 1.8 mm long, black; trochanter 0.6 mm long, black; femur 3.2 mm long, black; tibia 2.5 mm long, black, with spur; tarsus 2 mm long, black; pre-tarsus 1.1 mm long (4 – segmented), black coloured.

Mid Leg

10.7 mm long, black; coxa 1.1 mm long, black; trochanter 0.4 mm long, black; femur 2.7 mm long, black; tibia 3.1 mm long, black; tarsus 2.2 mm long, black; pre-tarsus 1.3 mm long (4 – segmented), black coloured.

Hind Leg

16 mm long, black; coxa 1.4 mm long, black; trochanter 0.4 mm long, black; femur 4 mm long, black; tibia 5 mm long, pale black; tarsus 3 mm long, black; pre-tarsus 2.2 mm long (4 – segmented), black coloured.

Abdomen

Abdomen 1.2 mm long, 1 mm broad, short, globose, black; petiole (abdominal segment 2), 0.1 mm long, node of pedicel broad, biconvex, armed with four short sub equal spines placed about equidistant from each other; tergite of first gastral segment large, half the length of gaster in dorsal view, first tergite longer than second; spines present on pronotum, propodeum and petiole.

Sting – Absent.

Colour

Blackish: Head, antenna, thorax, legs, abdomen.

Black: Eyes.

Host: Scales, Mealybugs.

Host plant – *Saccharum officinarum* L.

Paratype: 4 workers, Coll. Kurane, S.H. Mar. 2012 to Nov. 2014,head, antenna, leg, abdomen mounted on the card sheet and labeled as above.

Distributional Record

1 worker Malakapur, 17-III-2012; 2 Kolhapur, 11-VI-2012; 1 Shirala, 4-V-2013.

Polyrachis indica sp.n.

(Plate 18, Figures 72 to 76)

Worker (Figure 62 and 63)

Body 7 mm long; head 1.1 mm long, 1 mm broad; antenna 12 segmented; thorax 2 mm long; hind leg 21.4 mm long; abdomen 3.9mm long, 2 mm broad.

Head (Figure 64)

1.1 mm long, 1 mm broad, silky yellowish, cephalic index 90.90, oval, narrowed posteriorly; mandibles triangular, 1 mm long, mandibular index 17.85; clypeus broad from front to back; frontal carinae present; eyes 1 mm long, large, situated behind midlength of sides, black.

Antenna (Figure 65)

12.5 mm long, silky yellowish, 12 segmented including scape; scape 6 mm long, 0.4 mm broad, silky yellowish, scape index 600; pedicel 1 mm long, 0.4 mm broad, silky yellowish; flagellum 5.5 mm long, 10 segmented, apical segments of antennal funiculus not forming a club.

Flagellar Formula

1 L/W = 2, 3 L/W = 1.5, 4 L/W = 1.5, A = 1.6

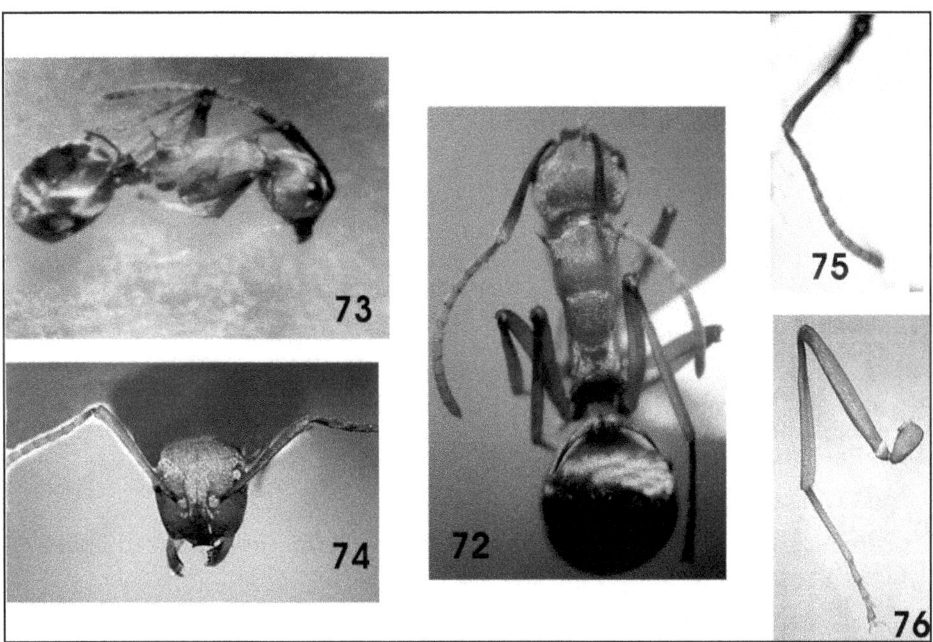

Plate 18: *Polyrachis indica* sp.n. Figure 72: Dorsal view; Figure 73: Lateral view; Figure 74: Head; Figure 75: Antenna; Figure 76: Hind leg.

Thorax

2 mm long,silky yellowish, longer than the head, pronotal and metanotal spines long, pronotal spines pointing forwards, metanotal and pedicel spines pointing backwards.

Wings – Absent.

Fore Leg

15.3 mm long, silky yellowish; coxa 2.2 mm long,silky yellowish; trochanter 0.5 mm long, silky yellowish; femur 4.3 mm long, silky yellowish; tibia silky yellowish; 3.5 mm long with spur; tarsus 2.5 mm long, silky yellowish; pre-tarsus 2.3 mm long (4- segments).

Mid Leg

15.7 mm long, silky yellowish; coxa 1 mm long, silky yellowish; trochanter 0.5 mm long, silky yellowish; femur 6 mm long,silky yellowish; tibia 4 mm long, silky yellowish; tarsus 2 mm long, silky yellowish; pre-tarsus 2.2 mm long (4 – segments).

Hind Leg (Figure 66)

21.4 mm long, silky yellowish; coxa 1.4 mm long, silky yellowish; trochanter 0.5 mm long, silky yellowish; femur 6.5 mm long, silky yellowish; tibia 6 mm long, silky yellowish; tarsus 4 mm long, silky yellowish; pre-tarsus 3 mm long (4 – segments).

Abdomen

Abdomen 3.9 mm long, 2 mm broad, 6 segmented,silky yellowish; petiole (abdominal segment 2) with pair of spines, pointing backwards, 1.1 mm long; first gastral tergite large half the length of gaster, first gastral tergite much longer than the second.

Sting – Absent.

Colour

Silky yellowish - Head, eyes, antenna, thorax, legs, abdomen.

Host plant – *Musa acuminata* Colla, 1820.

Paratype – 11 workers, Coll. Kurane, S.H. July. 2012 to Dec.2014.,head, antenna, leg and abdomen mounted on card sheet and labeled as above.

Distributional Record

11 workers Karad, 1-VII-2012

Remarks: This species doesn't match any of the species described under the genus *Polyrachis* and shows extra characters.

1. Body colour
2. Length of spines on thorax
3. Hence new species.

Etymology

This species reported first time from India, hence the name indica.

Sub-family Ponerinae

The sub family contains 6 tribes and 44 genera.

Tribe Ponerini

It contains 23 genera including *Harpegnathos* Jerdon.

Genus – *Harpegnathos* Jerdon, 1834

According to Bolton (2012) 7 species have been reported from world. In India 2 species have been reported. The genus shows following characters:

1. Head rectangular.
2. Mandibles articulated at the sides of the head and more than half as long as head.
3. Mandibles sickle- shaped curved up words.
4. Saw like teeth.
5. Clypeus triangular.
6. Antennal carinae stout.
7. Antennae 12- jointed, slender, filiform.
8. Eyes large, prominent.
9. Thorax elongate, depressed and laterally compressed.
10. Legs longer and slender.
11. Pedicel one- jointed, cylindrical, narrowed anteriorly.
12. Abdomen cylindrical.
13. Constriction between basal two segments, distinct.
14. Sting exerted, powerful.

Key to Species of the Genus *Harpegnathos*

Head, thorax and abdomen not concolorous.. *saltator*

Head, thorax and abdomen concolorous.. *venator*

***Herpegnathos saltator* Jerdon, 1851**

(Plate 19, Figures 77 to 81)

Worker (Figure 77 and 78)

Body 14 mm long; head 3 mm long, 2 mm broad; antenna 12 segmented; thorax 5 mm long; hind leg 30.9 mm long; abdomen 6 mm long, 2 mm broad; errect pale hairs on body; head, thorax and abdomen not concolorous.

Head (Figure 79)

3 mm long, 2 mm broad,ferruginous red, cephalic index 66.66, granulate; mandibles 13.5 mm long, mandibular index 170.88, forcep like; eyes 3 mm long, black; clypeus broad from front to back; frontal lobes present.

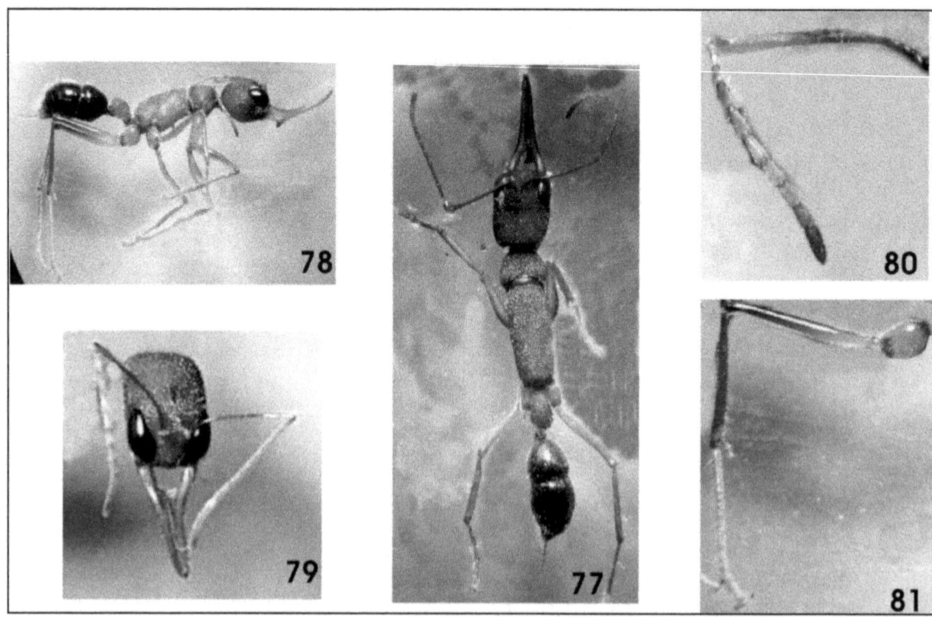

Plate 19: *Herpegnathos saltator*. Figure 77: Dorsal view; Figure 78: Lateral view; Figure 79: Head; Figure 80: Antenna; Figure 81: Hind leg.

Antenna (Figure 80)

17.3 mm long, yellow, 12 segmented including scape; scape 7.3 mm long, 0.4 mm broad, yellow, scape index 158.69; pedicel 1 mm long, 0.4 mm broad; flagellum 9 mm long, 10 segmented.

Flagellar Formula

1 L/W = 2.5, 3 L/W = 3.3, 4 L/W = 3, A = 2.9

Thorax

5 mm long, ferruginous red, granulate, covered with short errect pale hairs.

Wings – Absent.

Fore Leg

25.6 mm long, yellow; coxa 4.4 mm long,yellow; trochanter 1 mm long, yellow; femur 7.5 mm long, yellow; tibia 6 mm long, yellow, tibia with spur; tarsus 3.7 mm long, yellow; pre-tarsus 3 mm long (4 – segmented).

Mid Leg

20.4 mm long, yellow; coxa 2.1 mm long, yellow; trochanter 1 mm long, yellow; femur 6 mm long, yellow; tibia 5.3 mm long, yellow; tarsus 3 mm long, yellow; pre-tarsus 3 mm long (4 – segmented).

Hind Leg (Figure 81)

30.9 mm long, yellow, yellow; coxa 2.8 mm long, yellow; trochanter 1 mm long, yellow; femur 8.5 mm long, yellow; tibia 8.6 mm long, tibia with spur, yellow; tarsus 6 mm long, yellow; pre-tarsus – 4 mm long, yellow (4 – segmented).

Abdomen

Abdomen 6 mm long, 2 mm broad, black, shining, covered with short errect pale hairs; the petiole (abdominal segment 2) 1 mm long, narrowly attached to first gastral segment, constriction between abdominal segment 3 and 4 (gastral segment 1 and 2).

Sting

Present, large, strongly developed.

Colour

Ferruginous red: Head, thorax.

Yellow: Antenna, leg.

Black: Abdomen, eyes.

Host: Unknown

Host plant- Unknown

Paratype: 2 workers, Coll. Kurane, S.H. Mar. 2012 to Dec.2014, head, antenna, leg, abdomen mounted on the card sheet and labeled as above.

Distributional Record

1 worker Radhanagari, 4-III-2013; 1 Amba, 6-IV-2012.

Genus – *Leptogenys* Roger, 1861

According to Bolton (2012) the genus *Leptogenys* contains 307 species from the world and genus shows following characters:

1. Head quadrangular, as broad as long, broader in front than posterior.
2. Mandibles long, slender and curved.
3. Clypeus narrow.
4. Antennal carinae small.
5. Antenna 12- segmented, long, filiform.
6. Eyes large;
7. Thorax narrower than head.
8. Legs long.
9. Pedicel one-jointed, node cubical.
10. Abdomen cylindrical, constriction between the basal two segments.
11. Sting long.

Key to Species of the Genus *Leptogenys*

1. Petiolar node squamiform, compressed 2
   Petiolar node nodiform, not compressed 4
2. Clypeus with 3 teeth .... *dentilobis*
   Clypeus unarmed 3
3. Head subquaderate .... *birmana*
   Head rectangular .... *processiinalis*
4. Head striate 5
   Head punctured or smooth and shining never striate 12
5. Head entirely striate 6
   Head without striation 8
6. Head longitudinally striate 8
   Head longitudinally striate, pronotum and mesonotum rugose 7
7. Head sub rectangular .... *diminuta*
   Head oval .... *diminuta deceptrix*
8. HW > 1.55 .... *kitteli*
   Smaller species, HW < 1.55 .... *kitteli minor*
9. Body yellowish, head oval .... *diminuta woodmasoni*
   Body black, head subrectangular 10
10. Clypeus carinate 11
    Clypeus not carinate .... *diminuta palliseri*
11. Vertex smooth, shiny without striation .... *diminuta laeviceps*
    Vertex with feeble striation .... *diminuta diminuta laeviceps*
12. First gastral segment opaque 13
    First gastral segment smooth 15
13. HW > 1.45 mm .... *binghamii*
    HW < 0.95 mm 14
14. Petiolar node broader than long in dorsal view .... *hysterica*
    Petiolar node longer than broad .... *punctiventris*
15. Petiolar node as broad as long in dorsal view; petiolar dorsum broadly rounded in lateral view 16
    Petiolar node longer than broad in dorsal view, dorsum sloping anteriorly in lateral view 21
16. Cephalic dorsum smooth and shining 17
    Cephalic dorsum opaque 18

17. Second antennal segment two- fifths longer than third .... *lucidula*
    Second and third antennal segments sub equal .... *emiliae*
18. Mandibular basal margin toothed 19
    Mandibular basal margin not toothed 20
19. Head rectangular, with a blue metallic tint ....*moelleri*
    Head sub quadrate, with no blue metallic tint ....*dalyi*
20. Gaster lighter than head and mesosoma ....*roberti*
    Gaster unicolorous with head and mesosoma .... *roberti coonoorensis*
21. Cephalic dorsum punctate 22
    Cephalic dorsum smooth and shinning 24
22. HW > 0.95 mm; petiolar node broader than long ....*jeanettei*
    HW < 0.95 mm; petiolar node longer than broad 23
23. Body black..*lattkei*
    Body brown ....*transitionis*
24. Body brown ....*assamensis*
    Body black, third antennal segment shorter than second..................................25
25. Anterior clypeal margin sinusoid..*chinensis*
    Anterior clypeal margin straight ....*peuqueti*

Leptogenys chinensis **Mayr, 1870**

(Plate 20, Figures 82 to 86)

Worker (Figure 82 and 83)

Body 6.5 mm long, black; head 1 mm long, 0.9 mm broad; antenna 12 segmented: thorax 2.3 mm long; hind leg 19.3 mm long; abdomen 3.2 mm long, 1.1 mm broad.

Head (Figure 84)

1 mm long, 0.9 mm broad, black; cephalic index 90; eyes large, 0.7 mm long, black; mandibles triangular, 1.3 mm long, mandibular index 130; clypeus blackish coloured, anterior margin sinusoid; cephalic dorsum smooth and shining.

Antenna (Figure 85)

10.6 mm long, ferruginous yellow,12 segmented including scape; scape 4.1 mm long,0.4 mm broad, scape index 488; pedicel 0.5 mm long, 0.3 mm broad, ferruginous yellow; flagellum 6 mm long, 10 segmented, ferruginous yellow, third antennal segment shorter than second.

Flagellar Formula

1 L/W = 3, 3L/W=2.3, 4 L/W = 2, A= 2.4

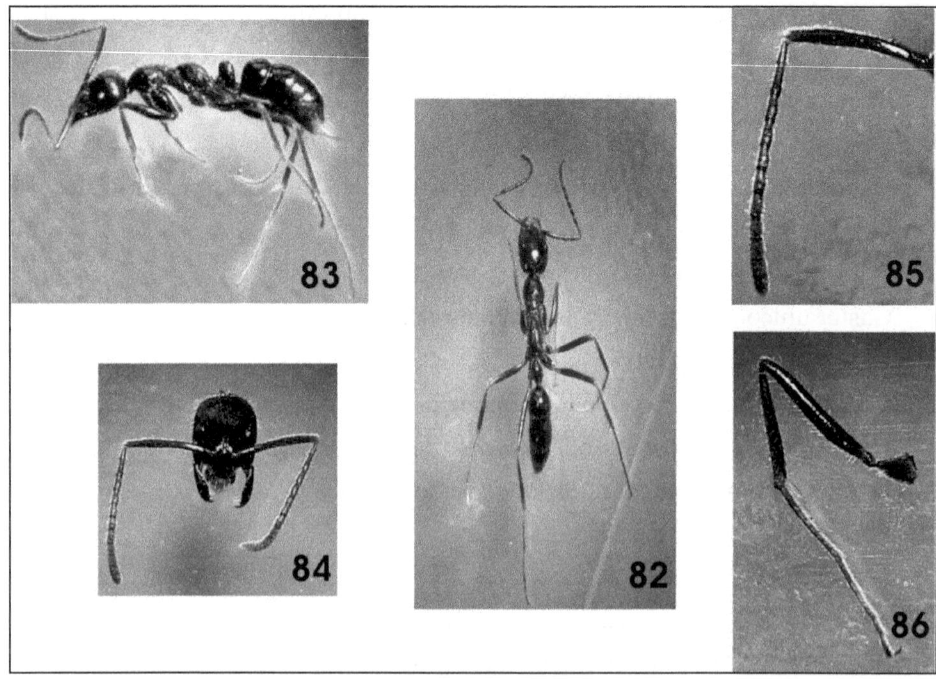

Plate 20: *Leptogenys chinensis*. Figure 82 Dorsal view; Figure 83: Lateral view; Figure 84: Head; Figure 85: Antenna; Figure 86: Hind leg.

Thorax

2.3 mm long, black; pronotum convex.

Fore Leg

15 mm long, ferruginous yellow; coxa 2.5 mm long, ferruginous yellow; trochanter 0.7 mm long, ferruginous yellow; femur 4.3 mm long, ferruginous yellow; tibia 4 mm long, with spur, ferruginous yellow; tarsus 1.5 mm long, ferruginous yellow; pre –tarsus 2 mm long (4 – segmented).

Mid Leg

14.4 mm long, ferruginous yellow; coxa 1.5 mm long, ferruginous yellow; trochanter 0.5 mm long, ferruginous yellow; femur 3.4 mm long, ferruginous yellow; tibia 3.5 mm long, ferruginous yellow; tarsus 2.3 mm long, ferruginous yellow; pre –tarsus 3.2 mm long (4 – segmented).

Hind Leg (Figure 86)

19.3 mm long, ferruginous yellow; coxa 1.7 mm long, ferruginous yellow; trochanter 0.6 mm long, ferruginous yellow; femur 4.8 mm long, ferruginous yellow; tibia 4.3 mm long, ferruginous yellow; tarsus 3.5 mm long, ferruginous yellow; pre –tarsus 4.4 mm long (4 – segmented).

Abdomen

3.2 mm long, 1 mm broad; waist of 1 segment the petiole (abdominal segment 2), 1 mm long, narrowly attached to first gastral segment; constriction between abdominal segments 3-6; petiolar node longer than broad, dorsum slopping anteriorly in lateral view.

Sting – Large, strongly developed, pointed, ferruginous yellow.

Colour

Black: Head, thorax.

Ferruginous yellow: Antenna, legs, abdomen.

Host plant: *Zizyphus jujuba* Miller, 1768.

Host- Lepidopterous caterpillars.

Paratype- 7 workers, Coll. Kurane, S.H. Jan. 2012 to Dec.2014, head, antenna, leg, abdomen mounted on the card sheet and labeled as above.

Distributional Record

2 workers Gadhingalaj, 2-III-2012; 3 Radhanagari, 7-IV-2012; 2Ajara, 9-V-2013.

Sub-family Dolichoderinae

It contains 4 tribes and 28 genera reported from the world. Tribe Dolichoderini contains 24 genera including *Tapinoma*.

Genus – *Tapinoma* Forster, 1850

The genus *Tapinoma* currently comprises 69 described species distributed worldwide (Bolton, 2012) and 2 species have been reported from India. The genus shows following characters:

1. Mandibles triangular.
2. Clypeus broad.
3. Antennae 12- jointed, filiform.
4. Eyes large.
5. Thorax narrower than head.
6. Legs long and slender.
7. Pedicel with node flat, inclined to the front.
8. Abdomen oval.

Key to Species of the Genus *Tapinoma*

Antennae long, scape extending beyond the top of the head ....*melanocephalum*

Antenna short, scape not extending beyond the top of the head .... *indicum*

Tapinoma melanocephalum Fabricius, 1793
(Plate 21, Figures 87 to 91)

Worker (Figure 87 and 88)

Body 1.5 mm long; head 0.3 mm long, 0.2 mm broad; antenna 12 segmented; thorax 0.4 mm long; hind leg 2.1 mm long; abdomen 0.8 mm long, 0.7 mm broad.

Head (Figure 89)

0.3 mm long, 0.2 mm broad, brownish red, cephalic index 66.66; smooth, longer than broad, oval, rounded posteriorly; mandibles triangular, 0.1 mm long, broad, mandibular index 33.33; eyes 0.1mm long, black, comparatively large, placed forward; clypeus convex.

Antenna (Figure 90)

1.9 mm long, yellowish white,12 segmented including scape; scape 0.6 mm long, 0.1 mm broad, yellowish white, scape length greater than head length, scape index 300; pedicel 0.3 mm long, 0.1 mm broad; flagellum 1 mm long, 10 segmented.

Flagellar Formula

1 L/W =1.1, 3 L/W = 1, 4 L/W = 1, A= 1.03

Thorax

0.4 mm long, brownish red, not emarginated, pro-meso and meso-metanotal sutures distinct, slight constricted at the later suture, basal portion of the metanotum short.

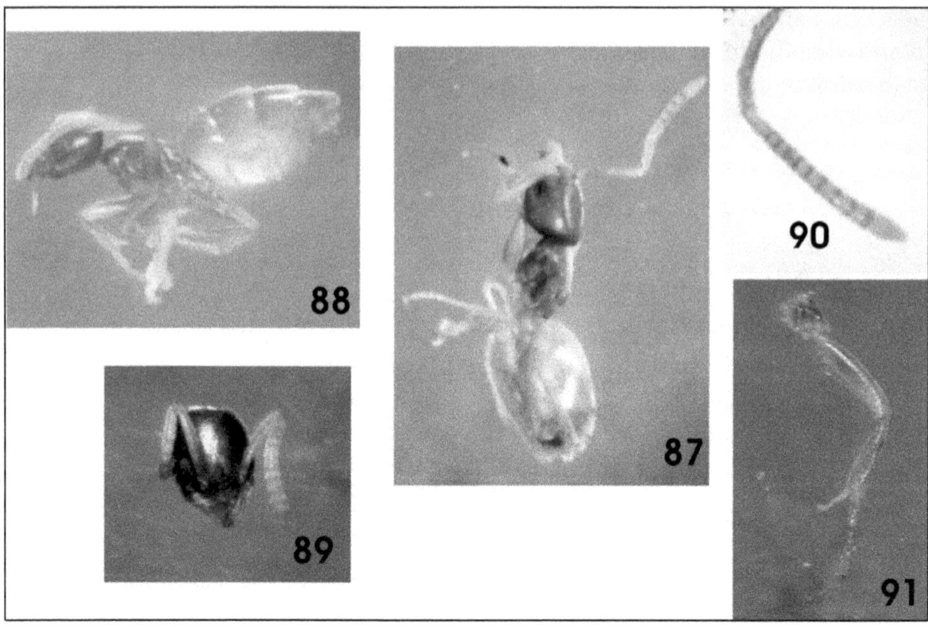

Plate 21: *Tapinoma melanocephalum*. Figure 87: Dorsal view; Figure 88: Lateral view; Figure 89: Head; Figure 90: Antenna; Figure 91: Hind leg.

Fore Leg

1.6 mm long, yellowish white; coxa 0.2 mm long, yellowish white; trochanter 0.1 mm long, yellowish white; femur 0.5 mm long, yellowish white; tibia 0.4 mm long, yellowish white; tarsus 0.1 mm long, yellowish white; pre-tarsus 0.3 mm long (4 – segmented), yellowish white coloured.

Mid Leg

1.2 mm long, yellowish white; coxa 0.1 mm long, yellowish white; trochanter 0.1 mm long, yellowish white; femur 0.4 mm long; yellowish white; tibia 0.3 mm long, yellowish white; tarsus 0.1 mm long, yellowish white; pre-tarsus 0.2 mm long (4 – segmented), yellowish white coloured.

Hind Leg (Figure 91)

2.1 mm long, yellowish white; coxa 0.3 mm long, yellowish white; trochanter 0.1 mm long, yellowish white; femur 0.7 mm long, yellowish white; tibia 0.5 mm long, yellowish white; tarsus 0.1 mm long, yellowish white; pre-tarsus 0.4 mm long (4 – segmented), yellowish white coloured.

Abdomen

Abdomen 0.8 mm long, 0.7 mm broad, yellowish white, elongate, oval; petiole (abdominal segment 2), 0.2 mm long, in profile usually a simple, overhung by first gastral segment and not visible in dorsal view; in dorsal view only 4 gastral tergite visible, fifth gastral tergite reflexed below the forth visible in ventral view where it forms a transverse plate abutting the fifth sternite; the anal and associated orifices are thus situated ventrally.

Sting – Absent.

Colour

Brownish red – Head, thorax

Yellowish white: Antenna, legs, abdomen.

Black: Eyes

Host plant: *Bambusa bambos* (Lin.)

Host- Bamboo aphids

Paratype: 9 workers, Coll. Kurane, S.H. Jan. 2012 to Dec.2014, head, antenna, leg, abdomen mounted on the card sheet and labeled as above.

Distributional Record

3 workers Miraj, 2-I-2012; 2 Sangli, 28-IX-2012; 2 Panhala, 3-X-2012; 2 Shirala, 19-XII-2013.

Sub-family Pseudomyrmecinae

It contains 1 tribe.

Tribe Pseudomyrmecini

It contains 3 genera including *Tetraponera* Smith.

Genus – *Tetraponera* Smith, 1852

The genus *Tetraponera* is characterized by their arboreal nature and slender bodied. 86 described species of *Tetraponera* are living in hollow structures of plants and trees, such as thorns or branches. *Tetraponera* species are closely related to the New World genus of ants *Pseudomyrmex*, but differ in their relationships with host plants. The genus shows following characters:

1. Antenna with 11- segments.
2. Antennal sockets well behind anterior margin of head.
3. Eyes large.
4. Metapleural gland orifice in lower posterior corner of metapleuron.
5. Pedicel two jointed (petiole and post petiole).
6. Sting present, usually large and strongly developed.

Key to Species of the Genus *Tetraponera*

It has been given by Bingham (1903).

Tetraponera rufonigra Jerdon, 1851

(Plate 22, Figures 92 to 96)

Worker (Figure 92 and 93)

Body 10 mm long; head 2 mm long, 1.1 mm broad; thorax 3 mm long; hind leg 19.3 mm long; abdomen 5 mm long, 1 mm broad.

Head (Figure 94)

2 mm long, 1.1 mm broad, black; cephalic index 55, eyes 1.7 mm long, reddish, very large and elongate at front of the mid length of the head; clypeus broad to reduced from front to back; mandibles reddish brown, 2 mm long, mandibular index 100; frontal lobes present, narrow vertical carina present between antennal socket.

Antenna (Figure 95)

8.9 mm long, reddish, 12 segmented including scape; scape 3.5 mm long, 0.4 mm broad, reddish, scape index 318.18; pedicel 0.7 mm long, 0.3 mm broad, reddish; flagellum 4.7 mm long, 10 segmented.

Flagellar Formula

1 L/W = 1.3, 3 L/W = 1, 4 L/W = 1, A= 1.1.

Thorax

3 mm long, reddish brown, first segment of mesosoma (pronotum) concealed to the second segment (mesonotum) by a flexible joint.

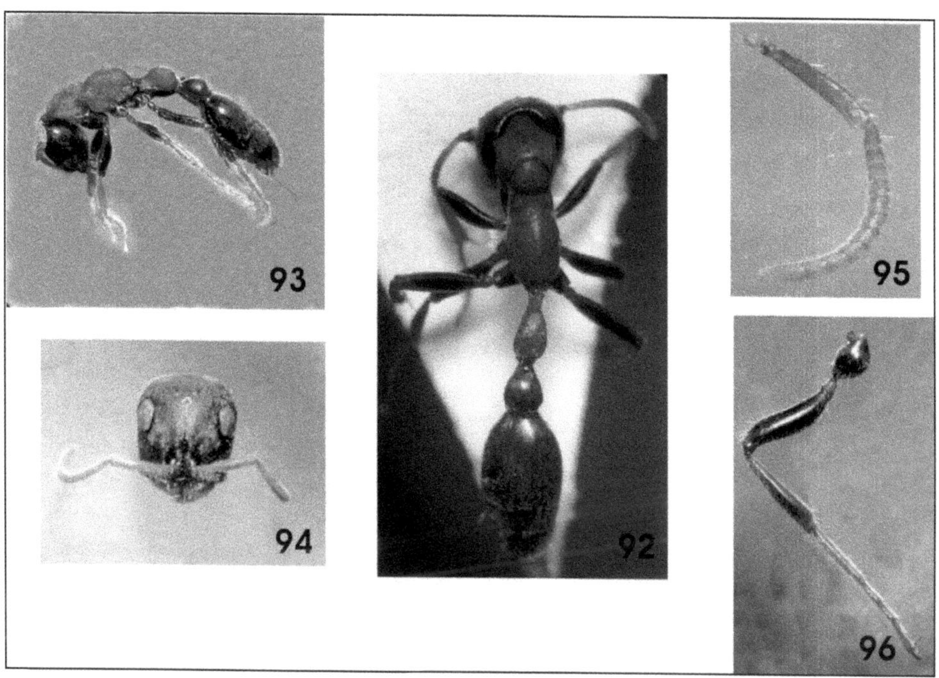

Plate 22: ***Tetraponera rufonigra.*** **Figure 92: Dorsal view; Figure 93: Lateral view; Figure 94: Head; Figure 95: Antenna; Figure 96: Hind leg.**

Wings – Absent.

Fore Leg

11.3 mm long, bi-coloured; coxa 2 mm long, black; trochanter 0.5 mm long, reddish; femur 3 mm long, black; tibia 2.5 mm long, black, with spur; tarsus 1.2 mm long, reddish; pre-tarsus 2.1 mm long, reddish.

Mid Leg

14.8 mm long, bi-coloured; coxa 1.6 mm long, black; trochanter 0.9 mm long, reddish; femur 3.9 mm long, black; tibia 4 mm long, black; tarsus 2.4 mm long, reddish; pre-tarsus 2 mm long, reddish.

Hind Leg (Figure 96)

19.3 mm long, bi-coloured; coxa 1.6 mm long, black; trochanter 1 mm long, reddish; femur 5 mm long, black; tibia 4.9 mm long, black, with spur; tarsus 3.8 mm long, reddish; pre-tarsus 3 mm long, reddish.

Abdomen

Abdomen 5 mm long, 1 mm broad, black; the petiole (abdominal segment -2), 1 mm long, reddish; the post petiole (abdominal segment -3), 0.5 mm long, black, post petiole attached to front lobe of gaster; pygidium large.

Sting – present, reddish, large and strongly developed.

Colour

Black: Head, eyes, coxa, femur, tibia, abdomen.

Reddish brown: Antenna, thorax, trochanter, tarsus, pre-tarsus, petiole, sting.

Host plant – Sugarcane *Saccharum officinarum* L.

Host – Scale insects.

Paratype: 9 workers, Coll. Kurane, S.H. Jan. 2012 to Dec.2014, head, antenna, leg, abdomen mounted on card sheet and labeled as above.

Distributional Record

3 workers Shirala, 25-I-2012; 4 Gadhingalaj, 20-II-2012; 2 Gaganbawda, 21-V-2013.

Sub-family Myrmicinae

It contains 6 tribes including Cataulacini and 141 genera.

Tribe Cataulacini

It contains 1 genus.

Genus – *Cataulacus* Smith, 1853

The genus contains 65 species reported from the world (Bolton, 2012) and 5 species have been reported from India. The genus shows following characters:

1. Head broad, flat, sides of the head deeply grooved to contain folded antennae.
2. Mandibles broad armed with4-5 teeth.
3. Frontal area triangular.
4. Antennal groove placed below the eyes.
5. Antennae short.
6. Scape and flagellum sub equal.
7. Antenna 11- segmented.
8. Thorax broad.
9. Legs short.
10. Tibiae flat.
11. Pedicel two-jointed (petiole and post- petiole).
12. Abdomen broadly oval.

Key to Species of Genus *Cataulacus*

a. Basal portion of metanotum with spines on posterior lateral angles

a1. Granular tubercles on abdomen, few on head and thorax

a2. Legs with tibiae orange-red above .... *taprobanae*

b2. Legs black .... *latus*

b1. Head, thorax and abdomen with granular tubercles

a2. First node of pedicel rounded ....*granulatus*

b2. First node of pedicel truncate ....*simoni*

b. Basal portion of metanotum without spines .... *muticus*

***Cataulacus taprobanae* Smith, 1853**

(Plate 23, Figures 97 to 101)

Worker (Figure 97 and 98)

Body 4.6 mm long, covered with a short, recumbent glistening gray pile; head 1.2 mm long, 1.2 mm broad; antenna 11 segmented; thorax 1.1 mm long; hind leg 8.3 mm long; abdomen 2.3 mm long, 1.4 mm broad.

Head (Figure 99)

1.2 mm long, 1.2 mm broad, intense black, white bristly hairs, as broad as long, sides of the head rounded; cephalic index 100; mandibles 0.6 mm long, mandibular index 50, stout, striate with a smooth, shining; clypeus triangular; eyes 0.9 mm, black, large, distinct and situated on the underside of the upper scrobe margin.

Antenna (Figure 100)

3.2 mm long,black and outer yellowish red,11segmented including scape; scape 1 mm long, 0.3 mm broad, yellowish red, scape index 83.33; pedicel 0.2 mm long, 0.3 mm broad, yellowish red; flagellum 3.2 mm long, 9 segmented, yellowish red; antennal scrobs present which run below the eye; apical segment of anterior funiculus not forming a club.

Flagellar Formula

1 L/W =0.5, 3 L/W = 0.5, 4 L/W = 0.5, A= 0.5

Thorax

1.1 mm long, black; white bristly hairs; pro-mesonotal shield convex, sides straight, narrowing and rounded posteriorly; meso-metanotal suture not distinct; basal portion of metanotum broader than long; lateral metanotal spines short, horizontal, directed backwards and slightly divergent.

Fore Leg

7.1 mm long,black and yellowish red; coxa 1 mm long, black; trochanter 0.2 mm long, black; femur 2.8 mm long, black; tibia 1 mm long, orange red above; tarsus 1.1 mm long, yellowish red; pre-tarsus 1 mm long, yellowish red (4 – segmented).

Mid Leg

6.2 mm long, black and yellowish red; coxa 1 mm long, black; trochanter 0.2 mm long, black; femur 2.1 mm long, black; tibia 0.9 mm long, yellowish red; tarsus 1 mm long, yellowish red; pre-tarsus 0.9 mm long, orange red (4 – segmented).

Plate 23: *Cataulacus taprobanae*. Figure 97: Dorsal view; Figure 98: Lateral view; Figure 99: Head; Figure 100: Antenna; Figure 101: Hind leg.

Hind Leg (Figure 101)

8.3 mm long, black and yellowish red; coxa 0.5 mm long, black; trochanter 0.3 mm long, black; femur 2.5mm long, black; tibia 1.5 mm long, orange red; tarsus 2 mm long, yellowish red; pre-tarsus 1.5 mm long, yellowish red (4 – segmented).

Abdomen

2.3 mm long, 1.4 mm broad, black; petiole (abdominal segment-2),0.4 mm long, black, short; post petiole (abdominal segment-3), 0.4 mm long, short, black; short, oval, convex, white bristly hairs

Sting – Absent

Colour

Black: Head, thorax, abdomen, eyes.

Yellowish red: scape and basal joint of flagellum, tibiae and tarsi of legs.

Host plant: *Mangifera indica* L.

Host- Jassids, Scales, Mealy bugs.

Paratype – 5 workers, Coll. Kurane, S.H. May. 2012 to Dec. 2014, head, antenna, leg, abdomen mounted on card sheet and labeled as above.

Distributional record-1 Malakapur, 2-V-2012; 2 Kolhapur, 11- VIII-2012; 2 Shirala, 23-XI-2014.

Tribe Crematogastrini

It contains 1 genus.

Genus - *Crematogaster* Lund, 1831

According to Bolton (2012) 494 species have been reported from the world and 29 species have been reported in India. *Crematogaster* is ecologically diverse genus of ants found worldwide, which is characterized by a distinctive heart-shaped gaster (abdomen), which gives one of their common names, Valentine ant. Members of this genus are also known as cocktail ants because of their habit of raising their abdomens when alarmed. Most of species are arboreal. These ants are sometimes known as acrobat ants. Acrobat ants acquire food largely through predation of other insects, like wasps. They use venom to stun their prey and a complex trail-laying process to lead comrades to food sources. Like many social insects, they reproduce in nuptial flights and the queen stores sperm as she starts a new nest. The genus shows following morphological characters,

1. Head square from front.
2. Mandibles strong and thick, narrow, with 4 teeth.
3. Antennae 11-jointed.
4. Club of the flagellum formed of 2,3or 4 joints.
5. Eyes lateral, moderate size.
6. Ocelli absent.
7. Thorax comparatively narrow and constricted at the junction of meso and metanotum.
8. First joint of pedicel broader, concave or flat above.
9. Second joint with rounded node.
10. Apex of pedicel attached to first abdominal segment.

Key to Species of the Genus *Crematogaster*

A. Metathorax not swollen;metanotum bispinous.

a. Head smooth and shining, at most with a striae anteriorly.

a1. Club of flagellum of antennae 4- jointed .... *wroughtoni*

b1. Club of flagellum of antennae 3-jointed.

a2. Pronotum sculptured.

a3. Pronotum convex, rounded in front.

a4. Lateral angles of pronotum prominent... *contemta*

b4. Lateral angles of pronotum not prominent.................................... *buddhce*

b3. Pronotum flat above, rounded in front.

a4. Eyes elongate .... *hodgsoni*

b4. Eyes round.

a5. Metanotal spines slender, not thick at base, divergent, straight, not curved .... *subnuda*

b3. Metanotal spines thick at base, less divergent, curved .... *anthracina*

b2. Pronotum smooth.

a3. Basal portion of metanotum sculptured.

a4. Pronotum with distinct lateral tubercle .... *sagei*

b4. Pronotum not tuberculate convex .... *walshi*

b3. Basal, portion of metanotum not sculptured, smooth.

a4. Head, viewed from the front, anteriorly truncate, raised into a high convex cone above .... *aberrans*

b4. Head, viewed from the front, anteriorly not truncate, not cone-shaped above.

a3. Pro-mesonotal suture obsolete slight.

a6. Metanotal spines short, much shorter than the length of the basal level portion of metanotum .... *politula*

b6. Metanotal spines long, longer than basal, level portion of metanotum.

a7. Cheeks and antennal hollows finely striate .... *travancorensis*

b7. Cheeks and antennal hollows not striate .... *ransonneti*

b5. Pro-mesonotal suture well marked, distinct.

a6. Mesonotum with a distinct transverse impression .... *dalyi*

b6. Mesonotum without any transverse impression.

a7. Cheeks finely striate .... *soror*

b7. Cheeks smooth, not striate .... *ebenina*

c1. Club of flagellum of antennae 2-jointed.

a2. Colour yellowish brown; 2nd joint of pedicel with a broad longitudinal groove above .... *millardi*

b2. Colour pale yellow; 2nd joint of pedicel with no longitudinal groove above .... *biroi*

b. Head not smooth, entirely sculptured

a1. Metanotal spines shorter than metanotum.

a2. Metanotal spines slender, apex directed backwards and outwards

b4. Eyes round.

a5. Metanotal spines slender, not thick at base, divergent, straight, not curved .... *subnuda*

a5. Metanotal spines thick at base, less divergent curved .... *anthracina*

b2. Pronotum not sculptured, smooth.

a3. Basal level portion of metanotum sculptured.

b4. Pronotum with distinct lateral tubercles .... *sagei*

b4. Pronotum not tuberculate, convex .... *walshi*

b3. Basal, level portion of metanotum not sculptured but smooth.

a4. Head, viewed from the front, anteriorly truncate, raised into a high convex cone above .... *aberrans*

b4. Head, viewed from the front, anteriorly not truncate, not cone-shaped above.

a5. Pro-mesonotal suture very slight.

a6. Metanotal spines short, much shorter than the length of the basal, level portion of metanotum .... *politula*

b6. Metanotal spines long, longer than basal, level portion of metanotum.

a7. Cheeks and antennal hollows finely striate .... *travancorensis*

b7. Cheeks and antennal hollows not striate.................................... *ransonneti*

b5. Pro-mesonotal suture well marked, distinct.

a6. Mesonotum with a distinct transverse impression............................. *dalyi*

b6. Mesonotum without any transverse impression.

a7. Cheeks finely striate.. *soror*

b7. Cheeks smooth, not striate.. *ebenina*

c1. Club of flagellum of antennae 2- jointed.

a2. Colour yellowish brown; 2nd joint of pedicel with a broad longitudinal groove above.. *millardi*

b2. Colour pale yellow; 2nd joint of pedicel with no longitudinal groove above.. *biroi*

b. Head not smooth, sculptured,

a1. Metanotal spines shorter than metanotum.

a2. Metanotal spines slender, apex directed backwards and outwards.

a3. First flattened joint of pedicel with the sides angular in the middle .... *dohrni*

b3. First flattened joint of pedicel with the sides not angular, rounded .... *artifex*

b2. Metanotal spines very thick at base, apex directed backwards and inwards.. *rothneyi*

b1. Metanotal spines distinctly longer than metanotum.

a2. Pronotum reticulate.

a3. First flattened joint of pedicel as broad as long, the sides angular in the middle .... *rogenhoferi*

b3. First flattened joint of pedicel distinctly longer than broad; sides nearly straights, lightly curved outwards .... *himalayana*

b2. Pronotum longitudinally striate.

a3. Mesonotum with a medial longitudinal carina .... *mogdiliani*

b3. Mesonotum not carinate in the middle.

a4. First flattened joint of pedicel, with the sides strongly arched, semicircular .... *flava*

b4. First flattened joint of pedicel with the sides straight, not arched .... *perekgans*

B. Metathorax remarkably broad, massive and swollen, no metanotal spines.

a. Metathorax yellow .... *inflata*

b. Metathorax black or dark castaneous brown.

a1. Meso-metanotal suture distinct, but not very deeply marked; base of metanotum above level with mesonotum .... *difformis*

b1. Meso-metanotal suture deep and broad; base of metanotum transversely raised and gibbous, higher than mesonotum .... *physothora*

Crematogaster rogenhoferi Mayr, 1878

(Plate 24, Figures 102 to 106)

Worker (Figure 102 and 103)

Body 3.5 mm long; head 0.8 mm long, 1.1 mm broad; antenna 11 segmented; thorax 1.2 mm long; hind leg 8.9 mm long; abdomen 1.5 mm long, 0.9 mm broad,pilosity pale.

Head (Figure 104)

0.8 mm long, 1.1 mm broad,reddish brown; square from front, cephalic index 137.5; eyes black, 0.4 mm long; mandibles 0.8 mm long, mandibular index 100; clypeus short; eyes 0.4 mm long, short, black.

Antenna (Figure 105)

4.3 mm long, thick, 11 segmented including scape, two segmented club, yellow; scape 2 mm long, 0.2 mm broad, yellow, reaching up to the top of head, scape index 250; pedicel 0.3 mm long, 0.2 mm broad, yellow; flagella 2 mm long, yellow, 10 segmented, 1st pedicel segment flattened and as broad as long.

Flagellar Formula

1L/W = 1, 3L/W = 1, 4L/W =1, A = 1.

Plate 24: *Crematogaster rogenhoferi*. Figure 102: Dorsal view; Figure 103: Lateral view; Figure 104: Head; Figure 105: Antenna; Figure 106: Hind leg.

Thorax

1.2 mm long, yellow, longitudinal; pronotum flat narrowing anteriorly; mesonotum narrower than mesonotum; mesonotum bears a long, slender spine, mesonotum smooth and shining.

Wings – Absent.

Fore Leg

5.4 mm long, yellow; coxa 0.6 mm long, yellow; trochanter 0.4 mm long, yellow; femur 2 mm long, yellow; tibia 1.6 mm long, with spur, yellow; tarsus 1 mm long, yellow; pre-tarsus 0.6 mm long, yellow (4 – segmented).

Mid Leg

7.5 mm long, yellow; coxa 0.5 mm long, yellow; trochanter 0.4 mm long; yellow; femur 2 mm long, yellow; tibia 2.6 mm long, yellow; tarsus 1.2 mm long, yellow; pre-tarsus 0.8 mm long (4 – segmented).

Hind Leg (Figure 106)

8.9 mm long, yellow; coxa 0.8 mm long, yellow; trochanter 0.4 mm long, yellow; femur 2.7 mm long, yellow; tibia 2.7 mm long, with spur, yellow; tarsus 1.6 mm long, yellow; pre-tarsus 1 mm long, yellow (4 – segmented).

Abdomen

1.5 mm long, 0.9 mm broad, reddish brown, elongate; pair of spine on propodeum; petiole (abdominal segment 2) 0.3mm long, yellow, low node, helcium; post –petiole, 0.1 mm long, post petiole articulated on dorsal surface of first gastral segment, gaster heart shaped; gastral first segment large.

Acidopore – Present.

Colour

Brown: Head.

Yellow: antenna, thorax, legs, petiole.

Host plant: *Tectona grandis* Linnaeus.

Host- Teak defoliator (Caterpillar)

Paratype: 9 workers, Coll. Kurane, S.H. Jan. 2012 to Dec.2014, head, antenna, hind leg, abdomen mounted on card sheet, labeled as above.

Distributional Record

3 workers Shahuwadi, 4-I-2012; 3 Shirala, 6-III-2012; 2 Walwa, 10-XI-2012; 1 Tasgaon, 5-III-2013.

***Myrmecaria sinuensis* sp.n.**

(Plate 25, Figures 107 to 111)

Worker (Figure 107 and 108)

Body 5 mm long; head 1 mm long, 1 mm broad; narrow neck; antenna 9.8 mm long; thorax 2 mm long; hind leg 24.4 mm long; abdomen 2 mm long, 1 mm broad; pilosity yellowish, fine and long, smooth and shining.

Head (Figure 109)

1 mm long, 1 mm broad, reddish yellow, much border than long, cephalic index 100; eyes present, 0.3 mm long, black; mandibles 0.4 mm long, triangular, rugose at base, mandibular index 28.57; clypeus broad.

Antenna (Figure 110)

9.8 mm long, reddish yellow, 7 segmented including scape; scape 3.2 mm long, 0.4 mm broad, reddish yellow, scape index 320; pedicel 0.7 mm long, 0.3 mm broad, reddish yellow; flagellum 5.9 mm long, 5 segmented.

Flagellar Formula

1 L/W = 3.3, 3 L/W = 3, 4 L/W = 2.25, A = 2.85.

Thorax

2 mm long, reddish yellow, narrow, constricted at the junction of the mesonotum and metanotum.

Wings – Absent.

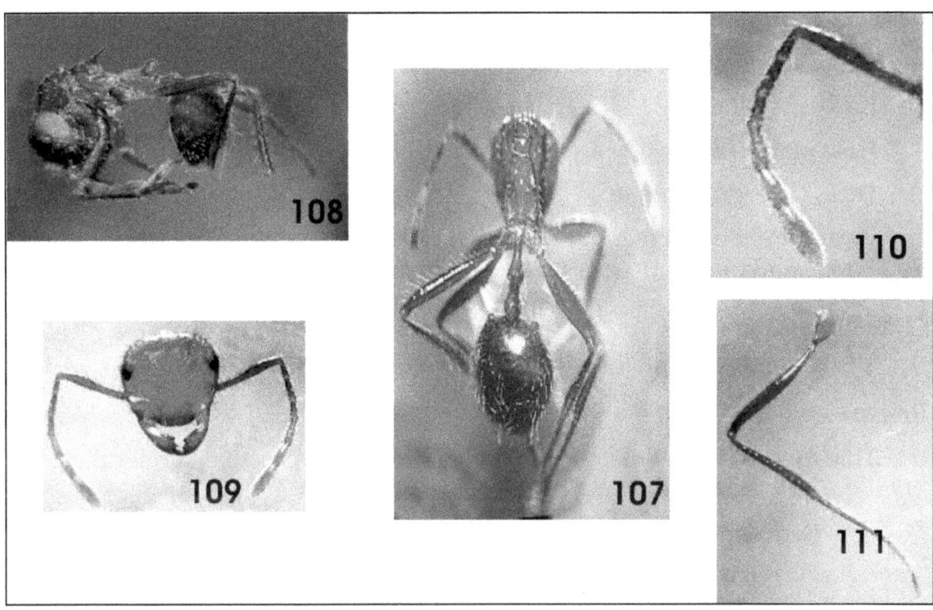

Plate 25: *Myrmecaria sinuensis* sp.n. Figure 107: Dorsal view; Figure 108: Lateral view; Figure 109: Head; Figure 110: Antenna; Figure 111: Hind leg.

Fore Leg

16.2 mm long, reddish yellow; coxa 1.9 mm long, reddish yellow; trochanter 0.5 mm long, reddish yellow; femur 4.8 mm long, reddish yellow; tibia 4 mm long, with spur, reddish yellow; tarsus 3 mm long, reddish yellow; pre-tarsus 2 mm long, reddish yellow (4 segmented).

Mid Leg

18.9 mm long, reddish yellow; coxa 1.4 mm long, reddish yellow; trochanter 0.5 mm long, reddish yellow; femur 6 mm long, reddish yellow; tibia 5 mm long, reddish yellow; tarsus 3 mm long, reddish yellow; pre-tarsus 3 mm long (4 – segmented).

Hind Leg (Figure 111)

24.2 mm long, reddish yellow; coxa 1.4 mm long, reddish yellow; trochanter 0.5 mm long, reddish yellow; femur 7 mm long, reddish yellow; tibia 6.4 mm long, reddish yellow; tarsus 5.4 mm long, reddish yellow; pre-tarsus 3.5 mm long (4 – segmented).

Abdomen

Abdomen 2 mm long, 1 mm broad, brownish, covered with erect pale hairs; petiole and post petiole (abdominal segment 2 and 3) petiolar length 0.5 mm long, reddish yellow, post petiole 0.5 mm long, reddish yellow, post petiole articulated on dorsal surface of first gastral segment, gaster heart shaped.

Colour

Reddish yellow: Head, antenna, thorax, legs.

Brown: Abdomen.

Host plant: *Tectona grandis* L.f.

Host- Caterpillars on teak.

Paratype: 08 workers, Coll. Kurane, S.H. Jan. 2012 to Dec.2014 head, antenna, leg abdomen mounted on card sheet and labeled as above.

Etymology

The abdomen resembles with the heart hence the name *Sinuensis.*

Remarks

This species doesn't match with any of the species described under the genus *Myrmecaria,* It shows following distinct characters

1. Body length very less
2. Flagellar formula
3. General body appearance

Therefore, this is new species.

***Crematogaster ashmeadi* Mayr, 1886**

(Plate 26, Figures 112 to 116)

Worker (Figure 112 and 113)

Body 2.1 mm long; head 0.2 mm long, 0.1 broad; antenna 11 segmented; thorax 0.9 mm long; hind leg 12.9 mm long; abdomen 1.1 mm long, 1.1 mm broad.

Head (Figure 114)

0.2 mm long, 0.1 broad, dark brownish red, cephalic index 150, smooth, with a few scattered erect hairs, longer than broad, oval, rounded posteriorly; mandibles 0.4 mm long, mandibular index 200, mandible triangular, broad, armed with numerous minute teeth; clypeus convex, broader than high; eyes 0.4 mm long, black.

Antenna (Figure 115)

8.3 mm long, yellowish white, 11 segmented including scape; scape 4.9 mm long, 0.2 mm broad, scape index 1633.3; pedicel 0.3 mm long, 0.1 mm broad, yellowish white; flagellum 3.1 mm long, 9 segmented.

Flagellar Formula

1 L/W = 2, 3 L/W = 3.5, 4 L/W = 3.5, A = 3.

Thorax

0.9 mm long, pale brownish, smooth, few scattered hairs, pro- meso and metanotal sutures distinct, slight constricted, basal portion of the metanotum short.

Plate 26: *Crematogaster ashmeadi.* Figure 112: Dorsal view; Figure 113: Lateral view; Figure 114: Head; Figure 115: Antenna; Figure 116: Hind leg.

Wings – Absent.

Fore Leg

11.5 mm long, yellowish white, coxa 1.5 mm long, yellowish white; trochanter 0.4 mm long, yellowish white; femur 3.6 mm long, yellowish white; tibia 3.1 mm long, yellowish white; tarsus 1.5 mm long, yellowish white; pre-tarsus 1.4 mm long,yellowish white (4-segmented).

Mid Leg

11.0 mm long, yellowish white; coxa 0.7 mm long, yellowish white; trochanter 0.4 mm long, yellowish white; femur 3.0 mm long, yellowish white; tibia 3.4 mm long, yellowish white; tarsus 2.0 mm long, yellowish white; pre –tarsus 1.5 mm long,yellowish white (4-segmented).

Hind Leg

12.9 mm long, yellowish white; coxa 0.8 mm long, yellowish white; trochanter 0.3 mm long, yellowish white; femur 3.8 mm long, yellowish white; tibia 3.8mm long, yellowish white; tarsus 2.5 mm long, yellowish white; pre –tarsus 1.7 mm long,yellowish white (4-segmented).

Abdomen

1.1 mm long, 1.1 mm broad, pale brownish, smooth, with a few scattered errect hairs, elongate, oval; waist of 1 segment; the petiole (abdominal segment 2), 0.1 mm long, yellowish white.

Sting – Absent.

Colour

Pale brownish red: head, abdomen.

Yellowish white: Antenna, thorax, legs.

Host plant: *Ficus racemosa* Linnaeus, 1753.

Paratype – 4 workers, Coll. Kurane, S.H. May. 2012 to Aug. 2014, head, antenna, leg, and abdomen mounted on card sheet and labeled as above.

Distributional Record

2 workers Kolhapur, 11-V-2012; 2 Shahuwadi; 5-VI-2012.

Myrmecaria kolhapurensis sp.n.

(Plate 27, Figures 117 to 121)

Worker (Figure 117 and 118)

4.5 mm long; head1.1 mm long, 1 mm broad; antenna 11.5 mm long; thorax 1 mm long; hind leg 24.4 mm long; abdomen 2.4 mm long, 1 mm broad; pilosity yellowish,few scattered whitish hairs, fine and long, smooth and shining.

Head (Figure 119)

1.1 mm long, 1 mm broad, dark chest nut-red, smooth,cephalic index 90.90; eyes present, 0.9 mm long, black, close to top of head; mandibles 1.4 mm long, triangular, rugose at base, mandibular index 127.27; clypeus rounded, convex in the middle.

Antenna (Figure 120)

11.5 mm long, chest nut-red,long, slender, 7 segmented including scape; scape 4.6 mm long, 0.4 mm broad, chest nut-red, scape index 460, scape reaching beyond the top of the head; pedicel 0.9 mm long, 0.2 mm broad, chest nut-red; flagellum 6mm long, 5 segmented, apical three segments formed a club out of which apical two much thickened.

Flagellar Formula

1 L/W = 2, 3 L/W = 2, 4 L/W = 2, A = 2.

Thorax

2 mm long, chestnut-red, narrow, pronotum large, rounded in front; pro-mesonotal suture distinct; mesonotum small; meso-metanotal suture well marked; basal portion of the metanotum rectangular; metanotal spines short and acute.

Wings – Absent

Fore Leg

16.2 mm long, chestnut-red; coxa 1.9 mm long, chestnut-red; trochanter 0.5 mm long, chestnut-red; femur 4.8 mm long, chestnut-red; tibia 4 mm long, with spur, chestnut-red; tarsus 3 mm long, chestnut-red; pre-tarsus 2 mm long, chestnut-red(4 segmented).

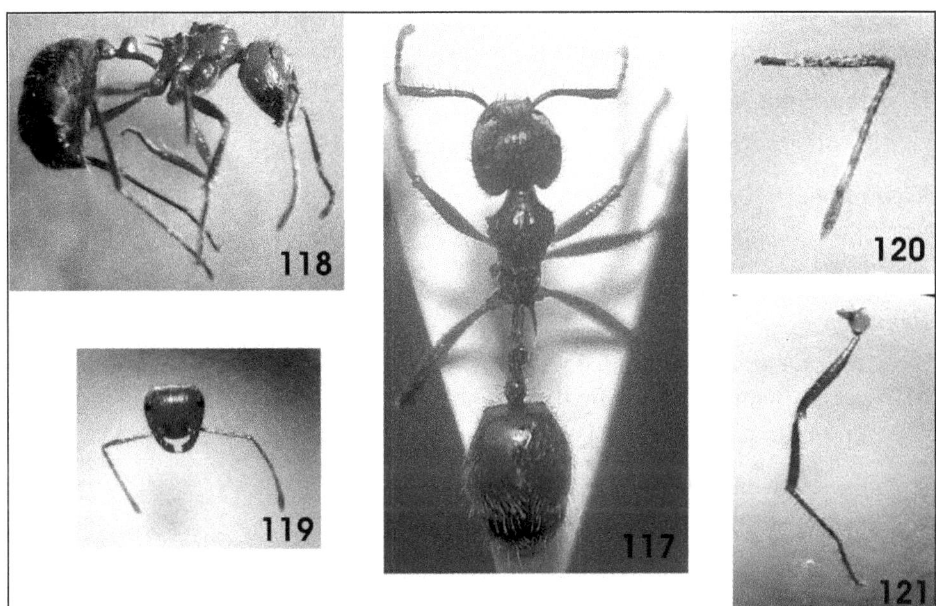

Plate 27: *Myrmecaria kolhapurensis* sp.n. Figure 117: Dorsal view; Figure 118: Lateral view; Figure 119: Head; Figure 120: Antenna; Figure 121: Hind leg.

Mid Leg

18.9 mm long, chestnut-red; coxa 1.4 mm long, chestnut-red; trochanter 0.5 mm long, chestnut-red; femur 6 mm long, chestnut-red; tibia 5 mm long, with spur, chestnut-red; tarsus 3 mm long,chestnut-red; pre-tarsus 3mm long, chestnut-red(4 segmented).

Hind Leg (Figure 121)

24.2 mm long, chestnut-red; coxa 1.4 mm long, chestnut-red; trochanter 0.5 mm long, chestnut-red; femur 7 mm long, chestnut-red; tibia 6.4 mm long, with spur, chestnut-red; tarsus 5.4 mm long, chestnut-red; pre-tarsus 3.5 mm long, chestnut-red (4 segmented).

Abdomen

2.4 mm long, black; whitish hairs; petiole (abdominal segment-2) 0.4 mm long, flat above not wide, semicircular in front; the post petiole (abdominal segment-3), 0.5 mm long, with transverse rounded tubercles at the apex.

Colour

Chestnut-red: Head, antenna, thorax, legs.

Black: Eyes, Abdomen.

Host plant: *Mangifera indica* L.

Host – Mealy bug, aphid.

Paratype: 04 workers, Coll. Kurane, S.H. Jan. 2012 to Dec.2014, head, antenna, leg abdomen mounted on card sheet and labeled as above.

Distributional Record

2 workers Amba, 4-I-2012; 2 Shahuwadi, 6-III-2012.

Etymology

This species reported first time from Kolhapur region hence the name *Kolhapurensis.*

Remarks

This species doesn't match with any of the species described under the genus *Myrmecaria,* It shows following distinct characters

1. Body lengthwise it is larger than other species
2. Flagellar formula
3. General body appearance.

Therefore, this is new species.

Tribe Solenopsidini

It contains 13 genera including *Solenopsis* Westwood.

Genus – *Solenopsis* Westwood, 1841

According to Bolton (2012), under this genus, 197 species have been reported from the world and 3 from India. The genus shows following characters:

1. Head square.
2. Mandibles thick and curved.
3. Clypeus narrow at the sides.
4. Antennal carinae short.
5. Antenna 10-jointed.
6. Club of the flagellum formed apical two joints.
7. Eyes small, rounded.
8. Thorax narrow.
9. Legs slender and long.
10. Pedicel two- jointed (petiole and post petiole).
11. Abdomen oval.

Key to Species of the Genus *Solenopsis*

a. Length over 3 mm..*geminata*

b. Length below 2 mm

a1. Reddish yellow, head and thorax opaque..*wroughtoni*

b1. Dark reddish brown, head and thorax smooth, shining.........................*nitens*

Solenopsis geminata Fabr, 1804

(Plate 28, Figures 122 to 126)

Worker (Figure 122 and 123)

Body 4.5 mm long, smooth, shining, polished; head 0.8 mm long, 0.9 mm broad; antenna 10 segmented; thorax 1.9 mm long; hind leg 9.8 mm long; abdomen 1.8 mm long, 0.8 mm broad.

Head (Figure 124)

0.8 mm long, 0.9 mm broad, reddish yellow, longer than broad, erect hairs; cephalic index 112.5; mandibles 0.5 mm long, mandibular index 62.5, borders are marked with brown; eyes black, minute, situated on underside of upper scrobe margin, anterior clypeal margin with a long, anteriorly projecting median seta.

Antenna (Figure 125)

4.2 mm long, 10 segmented including scape, with two segmented club, reddish yellow; scape 2.1 mm long, scape index 105, scape laid back in its normal resting position always distinctly above the eye, antennal scrobe absent, frontal carinae absent.

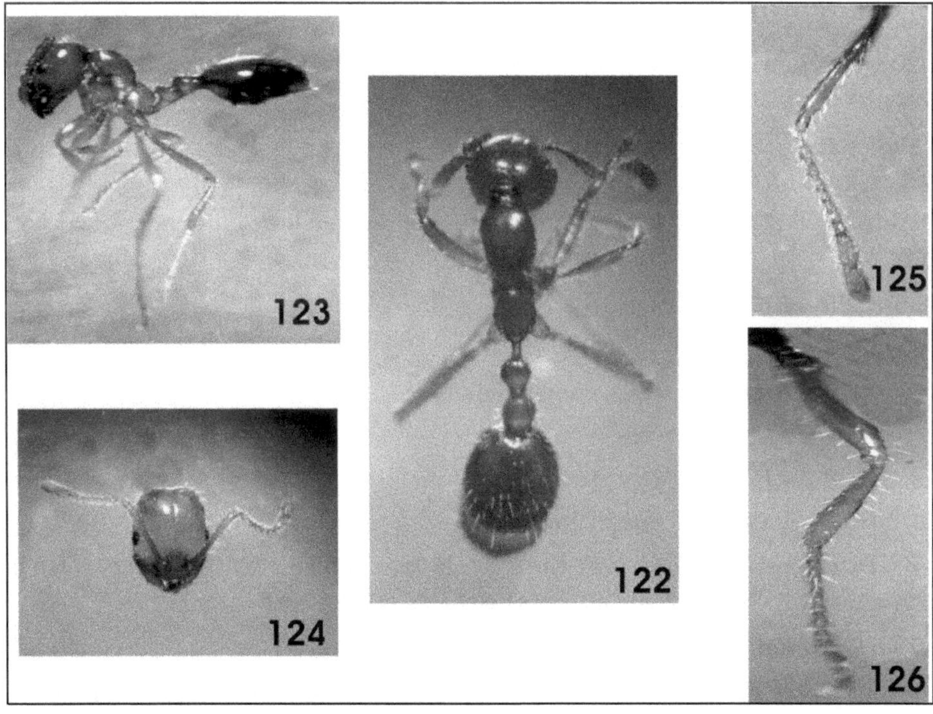

Plate 28: *Solenopsis geminata*. Figure 122: Dorsal view; Figure 123: Lateral view; Figure 124: Head; Figure 125: Antenna; Figure 126: Hind leg.

Flagellar Formula

1 L/W = 2, 3 L/W = 2, 4 L/W = 2, A= 2.

Thorax

1.9 mm long, reddish yellow, erect hairs.

Wings – Absent.

Fore Leg

7.9 mm long, reddish yellow; coxa 1 mm long, reddish yellow; trochanter 0.3 mm long, reddish yellow; femur 2.4 mm long, reddish yellow; tibia 1.9 mm long, with spur, reddish yellow; tarsus 1.4 mm long, reddish yellow; pre-tarsus 0.9 mm long, reddish yellow (4 – segmented).

Mid Leg

8.6 mm long, reddish yellow; coxa 0.9 mm long, reddish yellow; trochanter 0.5 mm long, reddish yellow; femur 3 mm long, reddish yellow; tibia 1.9 mm long, reddish yellow; tarsus 1.2 mm long, reddish yellow; pre-tarsus 1.1 mm long, reddish yellow (4 – segmented).

Hind Leg (Figure 126)

9.8 mm long, reddish yellow; coxa 0.9 mm long, reddish yellow; trochanter 0.4 mm long, reddish yellow; femur 1.9 mm long, reddish yellow; tibia 2.4 mm long, reddish yellow, with spur; tarsus 2.3 mm long, reddish yellow; pre-tarsus 1.9 mm long, reddish yellow (4 – segmented).

Abdomen

Abdomen 1.8 mm long, 0.8 mm broad, brownish; the petiole, 0.4 mm long, reddish yellow; post petiole 0.4 mm long, reddish yellow; borders marked with brown.

Sting – Present.

Colour

Reddish yellow – Head, antenna, thorax legs.

Brown – Abdomen.

Host plant – *Mangifera indica* L.

Host – Scale insect, mealy bug.

Paratype – 11workers, Coll. Kurane, S.H. May. 2012 to Dec. 2014, head, antenna, leg, abdomen mounted on card sheet and labeled as above.

Distributional Record

4 worker Sangli, 4-V-2012; 3 Palus, 11-VI-2012; 2 Shirala, 15-VII-2012; 2 Hatkanangale, 4-IX-2012.

Genus – *Monomorium* Mayr, 1855

The genus contains 396 species from the world (Bolton, 2012) and 17 species have been reported from India. It is considered to be "one of the more important genus of ants," considering its widespread distribution, its diversity, and its variety of morphological and biological characteristics. It also includes several familiar pest species, such as the pharaoh ant (*M. pharaonis*), the Singapore ant (*M. destructor*), and the flower ant (*M. floricola*). The genus shows following characters:

1. Head rectangular.
2. Mandibles narrow with 3 or 4 teeth.
3. Clypeus sub triangular.
4. Antennal carina short.
5. Antennae 11 or 12-jointed, club formed of apical three joints.
6. Eyes lateral, oval.
7. Thorax long and narrow.
8. Legs short or long and slender.
9. Pedicel two-jointed, first node short than second node.
10. Abdomen oval.

Key to Species of the Genus *Monomorium*

A. Head rugulose, opaque.

a. 2nd node distinctly broader than 1st node.

a1. Pro-mesonotum distinctly longer than broad

a2. Yellow, the abdomen posteriorly black

a3. Head posteriorly emarginate, length 1.5-2 mm .... *dichroum*

b3. Head posteriorly not emarginate, length 2.5-3mm .... *pharaonis*

b2. Head thorax and abdomen brown.

a3. Antennae long, scape reaching beyond top of head; sides of head convex .... *ongi*

b3. Antennae shorter, scape not attaining top of head; sides of head straighter, not convex

a4. Pro-mesonotum very convex, not margined at the sides .... *scharrn*

b4. Pro-mesonotum flatter, distinctly finely margined at the sides .... *wroughtoni*

b1. Pro-mesonotum short, as broad as long .... *fossulatum*

b. 2nd node not broader than 1st, nodes sub equal.

a1. Head in front distinctly broader than posteriorly .... *indicum*

b1. Head as broad posteriorly as in front .... *glyciphilum*

B. Head not rugulose and opaque, smooth and shining.

a. Head nearly square, almost as broad as long .... *aberrans*

b Head rectangular, distinctly longer than broad.

a1. Antennae 11 jointed.

a2 First node pedicel higher than 2nd node .... *orientale*

b2. First node of pedicel not higher than 2nd node .... *atomus*

b1. Antennae 12-jointed.

a2. Scape of antennae extending beyond top of head.

a3. Thorax convex above, not sub margined .... *sagei*

b3. Thorax flat, laterally sub margined .... *destructor*

b2. Scape of antennae not extending beyond top of head.

a3. Second node of pedicel not broader than 1st node.

a4. Head and thorax dark chest nut brown, abdomen black, length 1.5-2 mm .... *minutum*

b4. Head and thorax reddish yellow, abdomen dark brown, length 2.5-3 mm .... *gracillimum*

b2. Second node of pedicel broader than 1st node.

a1. Length 1.5-2 mm .... *floricola*

b4. Length 3-3.7 mm .... *latinode*

Monomorium minutum Buckely, 1866

(Plate 29, Figures 127 to 131)

Worker (Figure 127 and 128)

Body 2.6 mm long, Head 0.7 mm long, 0.6 mm broad; antenna 12 segmented; thorax 0.5 mm long; hind leg 8.1 mm long; abdomen 1.4 mm long, 0.5 mm broad.

Head (Figure 129)

Head 0.7 mm long, 0.6 mm broad, chestnut- brown, cephalic index 85.71; longer than broad, posteriorly transverse; mandibles narrow, 0.4 mm long, triangular, mandibular index 57.14; eyes 0.5 mm long, black, large, placed in middle of the sides of the head; clypeus convex, interiorly rounded.

Antenna (Figure 130)

4.9 mm long, chestnut- brown, moderately long,; 12 segmented including scape; scape 1.5 mm long, 0.2 mm broad, chestnut- brown, scape index 250, scape reaching up to the top of the head; pedicel 0.3 mm long, 0.1 mm broad, chestnut- brown; flagellum 3.1 mm long, 11 segmented.

Flagellar Formula

1L/W = 1, 3 L/W = 1, 4 L/W = 1, A = 1.

Thorax

0.5 mm long, chestnut- brown, pro-mesonotum convex, large.

Wings – Absent.

Fore Leg

7.7 mm long, chestnut brown; coxa 1.5 mm long, chestnut brown; trochanter 0.5 mm long, chestnut brown; femur 2.1 mm long, chestnut brown; tibia 1.6 mm long, cylindrical, chestnut brown, with spur; tarsus 0.7 mm long, chestnut brown; pre-tarsus 1.3 mm long, chestnut brown(4 – segmented).

Mid Leg

6.4 mm long, chestnut brown; coxa 1.1 mm long, chestnut brown; trochanter 0.5 mm long, chestnut brown; femur 2 mm long, chestnut brown; tibia 1.3 mm long, cylindrical, chestnut brown; tarsus 0.4 mm long, chestnut brown; pre-tarsus 1.1 mm long, chestnut brown(4 – segmented).

Hind Leg (Figure 131)

8.1 mm long, chestnut brown; coxa 1.1 mm long, chestnut brown; trochanter

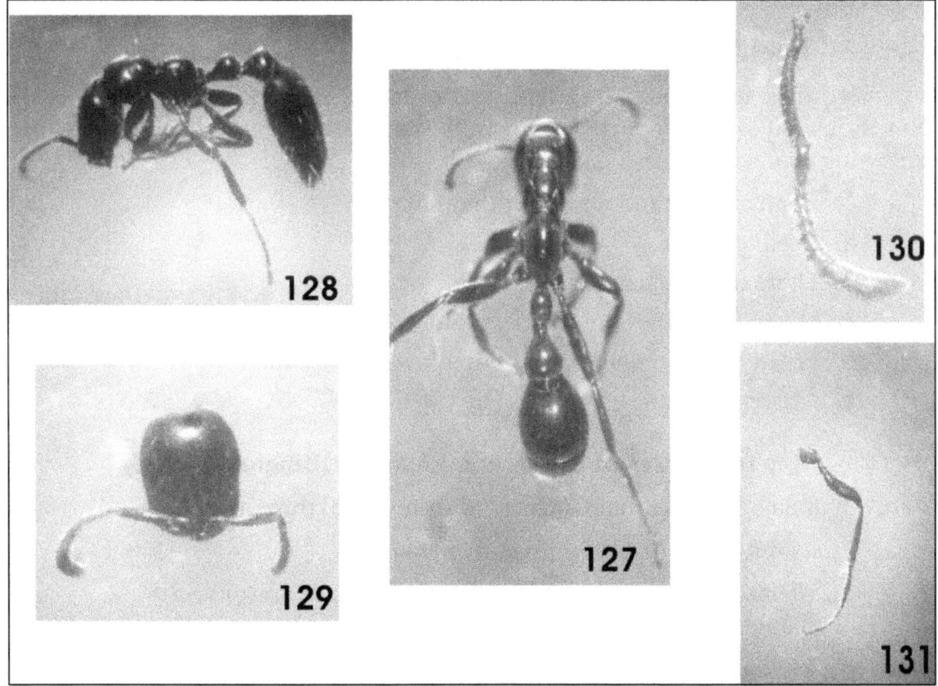

Plate 29: *Monomorium minutum*. Figure 127: Dorsal view; Figure 128: Lateral view; Figure 129: Head; Figure 130: Antenna; Figure 131: Hind leg.

0.5 mm long, chestnut brown; femur 2 mm long, chestnut brown; tibia 1.9 mm long, cylindrical, chestnut brown, with spur; tarsus 1 mm long, chestnut brown; pre-tarsus 1.6 mm long, chestnut brown (4 – segmented).

Abdomen

Abdomen 1.4 mm long, 0.5 mm broad, oval, black; the petiole (abdominal segment 2), 0.3 mm long,rounded; the post petiole (abdominal segment 3) 0.4 mm long, transverse, broader than long.

Colour

Chestnut brown - Head, thorax, antenna, legs.

Black –Abdomen

Host plant - *Tectona grandis* L.

Host – Lepidopterous caterpillar

Paratype – 9 workers, Coll. Kurane, S.H. April. 2012 to Oct. 2014, head, antenna, leg and abdomen mounted on card sheet and labeled as above.

Distributional Record

3 workers Shirala, 20-IV-2012; 3 Karad, 10-V-2012; 3 Kolhapur, 26-IV-2013.

Tribe Pheidolini

It contains 9 genera including *Pheidole* **Westwood.**

Genus – *Pheidole* Westwood, 1841

According to Bolton (2012), 1002 species have been reported from world and 50 species have been reported from India under this genus. The genus shows following characters:

1. Head smaller and narrower than thorax.
2. Ocelli present.
3. Thorax broad, flat above.
4. Metanotal spines present, triangular.
5. Pedicel two-jointed with nodes above.
6. Abdomen oval.

Key to Species of the Genus *Pheidole* (Bingham, 1903)

A. Club of flagellum of antennae formed from apical three joints.

a. First joint of the pedicel with a projection

a1. Metanotal spines clavate and obtuse towards apex like the halters or poisers of a dipteron .... *spathifera*

b1. Metanotal spines acute at apex, not clavate.

a2. Head posteriorly smooth and shining not sculptured .... *lamellinoda* b2. Head posteriorly sculptured.

a3. Frontal grooves for reception of scapes of antenna absent .... *grayi*

b3. Frontal grooves for reception of scapes of antennae distinct.

a4. Upper margin of node on 1st joint of pedicel emarginate .... *malinst*

b4. Upper margin of node on 1st joint of pedicel entire, not emarginate.

a5. Lateral lobes of head punctate .... *naorojii*

b5. Lateral lobes of head reticulate.

a6. Abdomen opaque, finely striate from end to end ....*sharpi*

b6. Abdomen not opaque, only striate at base.

a7. Second node distinctly more than twice as broad as long.

a8. Head, thorax and abdomen yellowish red...................................... *latinoda*

b8. Head, thorax and abdomen black or very dark brown...............*angustior*

b7. Second node of pedicel longer, less than twice as broad as long

a8. Head distinctly longer than broad...*fergusoni*

b8. Head square, as broad as long ....*hoogwerfi*

b. First joint of pedicel with no projection

a1. Pro- and mesonotum forming a single convexity, transverse mesonotal furrow obsolete.

a2. Head as broad as long.

a3. Posterior half of head smooth and shining .... *megacephala*

b3. Posterior half of head sculptured, not smooth.

a4. Occipital lobes longitudinally striate, length above 6mm ....*sykesi*

b6. Occipital lobes reticulate, length below 3 mm.....................................*sagei*

b2. Head distinctly longer than broad.

a3. Occiput smooth and shining.

a4. Occipital emargination narrow and deep, lateral lobes broad and rounded .... *pronotalis*

b4. Occipital emargination broad and shallow, lateral lobes narrow and angular .... *wood-masoni*

b3. Occiput sculptured.

a1. Length above 6 mm.

a3. Second node of pedicel with the sides produced into cone..........*phipsoni*

b5. Second node of pedicel with the sides not produced into cones...*hospita*

b4. Length below 4 mm.

a3. Head anteriorly beneath bi-dentate .... *watsoni*

b5. Head anteriorly beneath not dentate.

a6. Frontal grooves for reception of scapes of antennae absent .... *mus*

b6. Frontal grooves for reception of scapes of antennae present.

a7. Thorax for the most part smooth shining .... *templaria*

V. Thorax reticulate and opaque, not shining .... *parva*

b1. Pro- and mesonotum not forming a single convexity; transverse mesonotal furrow and ridge or carina, or at any rate the latter, always present.

a2. Posterior third of head smooth, not sculptured .... *nietneri*

b2. Whole head sculptured.

a3. Head below vertex vertically truncate, forming a flat plane with the clypeus .... *capellini*

b3. Head below vertex normally developed, not truncate.

a4. Mesonotum bi dentate above .... *multidens*

b1. Mesonotum not dentate above.

a5. Frontal grooves for reception of scape of antennae obsolete.

a6. Head including mandibles very large, as long as or longer than thorax and pedicel united...................................... *wroughtoni*

b6. Head including mandibles smaller, distinctly shorter than thorax and pedicel united .... *constanciae*

b5. Frontal grooves for reception of scape of antennae distinct.

a6. Length above 7 mm. scape flattened .... *rugosa*

b6. Length below 6 mm. Scape cylindrical.

a1. Vertex of head with a transverse impression, broad and very distinct.

a8. Abdomen finely striate at base only .... *sulcaticeps*

b8. Abdomen finely striate from end to end, opaque .... *yeensisp*

b7. Vertex of head not transversely impressed

a8. Pronotum convex, lateral tubercles obsolete.

a8. Base of abdomen finely striate, remainder of abdomen smooth and shining .... *fossulata*

b9. Abdomen entirely smooth and shining.

a10. Medial portion of clypeus smooth and shining.

a11. Frontal grooves not well-marked, coarsely longitudinally striate within .... *feae* b11. Frontal grooves more distinct, finely sculptured within .... *roberti*

b10. Medial portion of clypeus opaque, longitudinally striate.

a11. Scape of antennae comparatively long, falling short of apex of lateral lobe of head by about one-fifth of its own length .... *jucunda.*

b11. Scape of antennae shorter, falling: short of apex of lateral lobe of head by more than half its own length .... *jacana*

b8. Pronotum convex, lateral tubercles distinct.

a9. Second node of pedicel distinctly more than half the breadth of abdomen.

a10. Head longer than broad .... *horni*

b10. Head as broad as long .... *rhombinoda*

b9. Second node of pedicel not nearly half the breadth of abdomen.

a10. Clypeus medially produced bi dentate.. *peguensis*

b10. Clypeus not medially produced nor dentate.

a11. Abdomen sculptured.

a12. Basal third of abdomen finely striate .... *striativentris*

b12. Entire abdomen reticulate punctate, not striate .... *ghatica*

b11. Abdomen not sculptured, smooth and shining.

a12. Longitudinal striae on head curving outwards on posterior lateral lobes.

a13. Pronotum highly polished, smooth and shining.................... *sepulchralis*

b13. Pronotum more or less transversely striate.

b14. Frontal grooves for reception of scape longitudinally striate within .... *plagiaria*

a14. Frontal grooves for reception of scape finely reticulate within.

a15. Second node of pedicel transverse, with acute .... lateral cones... *binghamii*

a15. Second node of pedicel more rounded, the lateral angles more obtuse .... *indica*

b12. Longitudinal striae on head not curving outwards on posterior lateral lobes.

a13. Length below 3 mm .... *rogersi*

b13. Length above 3 mm.

a14. Head long, half as long again as broad... *magretti*

b14. Head shorter, about as long as broad.

a15. Metanotal spines long, longer than half the length of the basal face of metanotum .... *rotschana*

b15. Metanotal spines very short

a16. Pronotum seen from the front rounded, convex .... *himalayana*

b16. Pronotum seen from front flat anteriorly transverse above .... *allani*

B. Club of flagellum of antennae formed of apical four joints.

a. Light reddish-brown; head enormous, clypeus not carinate............ *smythiesi*

b. Very dark brown, almost black; head proportionately much smaller, clypeus medially carinate .... *bhavunae*

Pheidole megacephala Fabricius, 1793

(Plate 30, Figures 132 to 136)

Worker (Figure 132 and 133)

Body 4.8 mm long, shining, pilosity pale yellowish; head 1.8 mm long, 2 mm broad; antenna 5.5 mm long,12 segmented; thorax 1.2 mm long; hind leg 11.3 mm long;abdomen 1.8 mm long.

Head (Figure 134)

1.8 mm long, 2 mm broad, brownish red, large as compared to body; cephalic index 111; pilosity pale yellow; mandibles 1.3 mm long, mandibular index 72.22, with basal half flat and narrow and boarding towards masticatory margin; median portion of clypeus not vertical; eyes 0.5 mm long, black, present infront of the midlength of the sides; posterior half of the head smooth and shining.

Antenna (Figure 135)

5.5 mm long, brownish red, long, slender, 12 segmented including scape; scape 2 mm long, 0.2 mm broad,scape index 100, scape with 3- segmented club; pedicel 0.4 mm long, 0.2 mm broad; flagellum 3.1 mm long, 10 segmented.

Flagellar Formula

1L/W = 3, 3L/W = 2, 4 L/W = 3, A = 2.6.

Thorax

1.2 mm long, brownish red

Wings- Absent.

Fore Leg

8.4 mm long, brownish yellow; coxa 1 mm long, brownish yellow; trochanter 0.4 mm long, brownish yellow; femur 2.9 mm long, brownish yellow; tibia 2 mm long, brownish yellow; tarsus 1.1 mm long, brownish yellow; pre-tarsus 1 mm long, brownish yellow (4 – segmented).

Mid Leg

8.3 mm long, brownish yellow; coxa 0.5 mm long, brownish yellow; trochanter

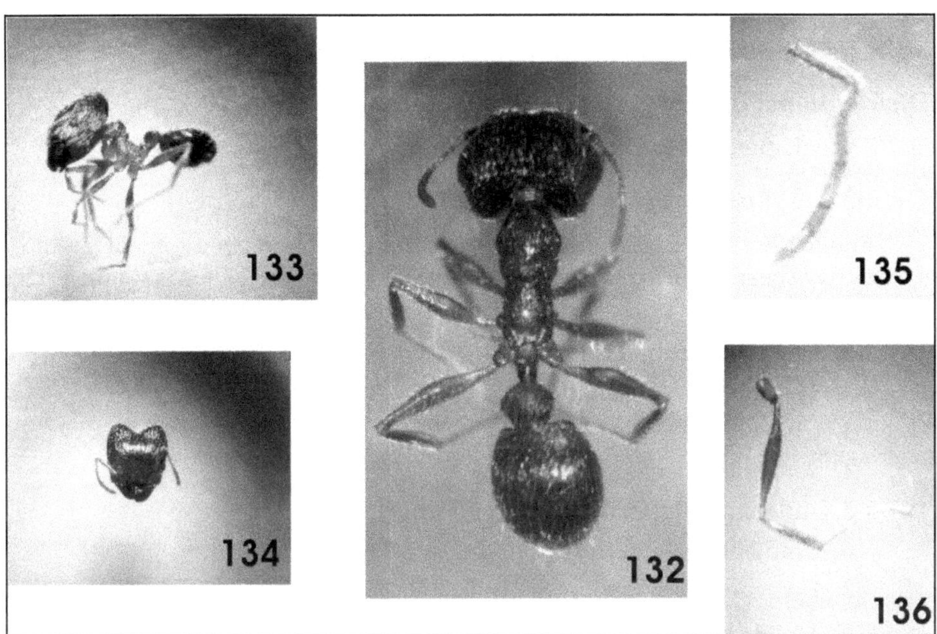

Plate 30: *Pheidole megacephala*. Figure 132: Dorsal view; Figure 133: Lateral view; Figure 134: Head; Figure 135: Antenna; Figure 136: Hind leg.

0.3 mm long, brownish yellow; femur 2.5mm long, brownish yellow; tibia 1.5 mm long, brownish yellow; tarsus 2 mm long, brownish yellow; pre-tarsus 1.5 mm long, brownish yellow (4 – segmented).

Hind Leg (Figure 136)

11.3 mm long, brownish yellow; coxa 0.6 mm long, brownish yellow; trochanter 0.4 mm long, brownish yellow; femur 3.5mm long, brownish yellow; tibia 2.2 mm long, brownish yellow; tarsus 3 mm long, brownish yellow; pre-tarsus 1.6 mm long, brownish yellow (4 – segmented).

Abdomen

Abdomen 1.8 mm long, 1 mm broad, dark brown, petiole(abdominal segment 2) brownish red, 0.3 mm long; post petiole (abdominal segment 3) brownish red, 0.4 mm long, articulated on anterior surface of first gastral segment.

Sting

Colour

Brownish red: Head, antenna, thorax.

Brownish yellow: Legs.

Dark brown: Abdomen.

Host plant: *Mangifera indica* L.

Host – Scale insect.

Paratype – 5 workers, Coll. Kurane, S.H. May. 2012 to Dec. 2014, head, antenna, leg, abdomen mounted on card sheet and labeled as above.

Distributional Record

2 Malakapur, 2-V-2012; 3 Kolhapur, 11- VIII-2012.

Genus – *Aphaenogaster* Mayr, 1853

About 186 species have been reported from the world (Bolton, 2012) and 8 species have been reported from India under this genus. The genus is characterized by:

1. Head longer than broad.
2. Mandibles sub triangular.
3. Clypeus flat.
4. Antennal carinae short.
5. Antenna 12-jointed, slender, filiform.
6. Club of flagellum formed of the apical four joints.
7. Eyes moderate size.
8. Thorax narrow, elongate.
9. Metanotal spines short.
10. Legs long and slender, tibiae with one spur.
11. Pedicel two-jointed, first node sub conical, second node oval longer than broader.

Key to Species of the Genus *Aphaenogaster*

A. Head posteriorly produced and constricted, forming a more or less cylindrical neck

 a. Metanotum transversely rugose above, striate on the sides................ *beccarii*

 b. Metanotum smooth

  a1. Length below 5 mm, second node of pedicel as broad as long.. *longipes*

  b1. Length above 5 mm, second node of pedicel as long as broad ....*feae*

B. Head not produced into neck

 a. Pronotum distinctly bituberculate.. *rothneyi*

 b. Pronotum not bituberculate

  a1. Head and thorax sparsely sculptured, shining

  a2. Pro-mesonotum forming convexity ....*sagei*

  b2. Pro-mesonotum not forming convexity

  a3. Antennae slender, joints 2-7 of flagellum three times as long as broad... *cristata*

  b3. Antennae robust, joints 2-7 of flagellum longer than broad....... *smythiesi*

***Aphaenogaster beccari* Emery, 1887**

(Plate 31, Figures 137 to 141)

Worker (Figure 137 and 138)

Body 6 mm long, shining; head 1 mm long, 0.5 mm broad; antenna 15 mm long, 12 segmented; thorax 2 mm long; hind leg 25.5 mm long, slender; abdomen 3 mm long, 1.1 mm broad.

Head (Figure 139)

1 mm long, 0.5 mm broad, castaneous brown,elongated, narrow; cephalic index 50, errect yellow hairs; mandibles 1 mm long, triangular, elongate, broad, mandibular index 100; median portion of clypeus rounded, extended backwards between the frontal carinae, eyes 0.8 mm long, black.

Antenna (Figure 140)

15 mm long, castaneous brown, very long, filiform, 12 segmented including scape, with four segmented club; scape 6mm long, 0.3 mm broad, scape index 1200; basal joint of flagellum longer than second segment, apical four segment longer.'

Flagellar Formula

1 L/W = 1.6, 3 L/W = 2, 4 L/W = 2, A = 1.8.

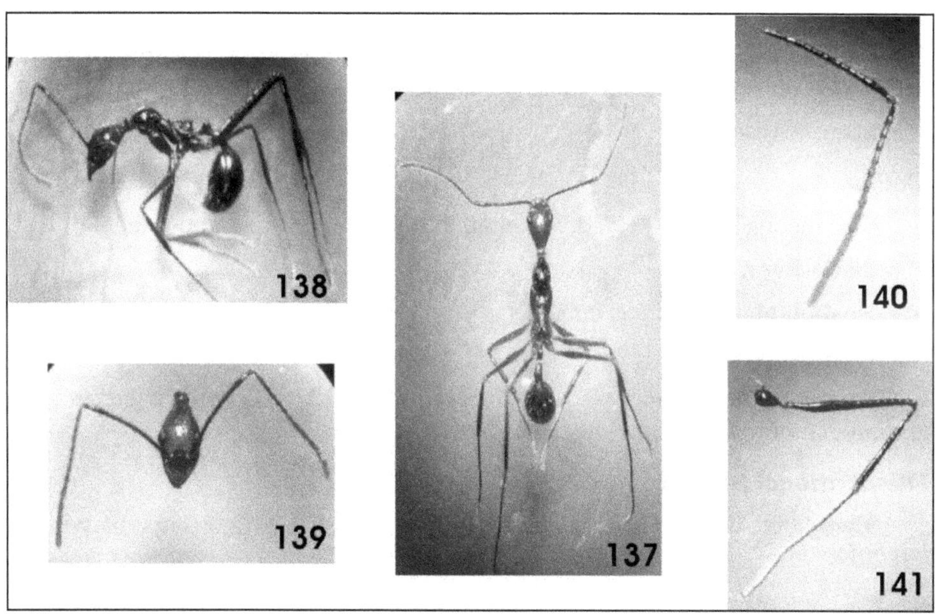

Plate 31: *Aphaenogaster beccari.* Figure 137: Dorsal view; Figure 138: Lateral view; Figure 139: Head; Figure 140: Antenna; Figure 141: Hind leg.

Thorax

2 mm long, castaneous brown darker, elongate, narrow; pronotum, mesonotum forming a high dome like arc; mesonotum convex, distinct, transversally rugose above, striate on the sides.

Wings – Absent.

Fore Leg

19.7 mm long, castaneous brown; coxa 2.7 mm long; trochanter 0.5 mm long; femur 6 mm long; tibia 4.5 mm long, tibia with spur; tarsus 4 mm long; pre-tarsus 2 mm long.(4-segmented).

Mid Leg

16.2 mm long, castaneous brown; coxa 1.4 mm long; trochanter 0.5 mm long; femur 4 mm long; tibia 4.4 mm long; tarsus 3.9 mm long; pre-tarsus 2 mm long. (4-segmented).

Hind Leg (Figure 141)

25.5 mm long, castaneous brown; coxa 1.5 mm long; trochanter 0.5 mm long; femur 8.5 mm long; tibia 7 mm long, spur absent; tarsus 5.5 mm long; pre-tarsus 2.5 mm long.(4-segmented).

Abdomen

Abdomen 3 mm long, 1.1 mm broad; castaneous brown, short, oval; petiole (abdominal segment 2) castaneous brown, 1 mm long, conical; post petiole (abdominal segment 3) 1 mm long, castaneous brown, oval, rounded above, articulated on anterior surface of first gastral segment.

Sting – Absent.

Colour

Castaneous brown – Head, antenna, thorax dark, leg and abdomen.

Black- Eyes.

Host plant – *Artocarpus integrifolia.*

Host - Unknown.

Paratype – 7 workers, Coll. Kurane, S.H. June. 201 2 to Dec.2014, head, antenna, abdomen mounted on card sheet and labeled as above.

Distributional Record

3 workers Ajara, 25-VI-2012; 2 Gadhingalaj, 20-VII-2012; 2Gaganbawda, 21-VII-2013.

5
Seasonal Abundance

Introduction

Ants (Hymenoptera: Formicidae) are social insects and found in all terrestrial habitats. Their morphology and size varied as per the habitat. They are docile to extremely pugnacious.Ants feed on plant seed, nector, honey dew secreted by cell sap sucking insects and fungi. They may be carnivores, scavengers, necrophages, predators and pests on economically important crop plants. They play an important role in food web and plant protection. Therefore, their seasonal abundance and distribution have been studied from both agro and forest ecosystems of Western Maharashtra especially, from the districts Kolhapur, Sangli and Satara. In past, ant abundance and their distribution have been studied by Agosti (1991), Arnold (1915), Bingham (1903), Bolton and Collingwood (1975), Wu (1990) *etc.* from the world. However, very little attension is given on ant abundance and distribution from Maharashtra except the work of Bhoje *et al.* (2013, 2014), Khot *et al.* (2013), Nagaria and Pawar (2014), Kurane *et al.* (2015), *etc.*

Materials and Methods

Ants were collected from different Tahasils of Kolhapur, Sangli and Satara (Figure 13) from both forest and agro ecosystems. The ants were collected in 70 per cent alcohol with the help of camel hair brush and stored temporarily in specimen tubes. The collected samples were identified in the laboratory by consulting taxonomic features of the species and literature cited under references. Occurrence and distribution was studied by visiting different study spots (Tahasils) of Kolhapur, Sangli and Satara (Figure 12) at 15 days interval and spot observations/collection by one man one hour search method.

Results

Results are recorded in Table 1 and Figures 142 to 160. The nesting habitats of some ants are shown in Figures 142 to152. Some ants like *Camponotus, Polyrachis, Tapinoma, Crematogaster* etc. were pests of crops like *A. esculentus, F. racemosa, Z. mays, S. bicolor, C. cajan etc.* while, some species were predatory upon certain small insects like midges, jassids, scales, whiteflies etc. There was commonsalization of ants with some insect pests like whiteflies, mealy bugs, scales, aphids and other homopterous insects. These insects secreted honeydew like substance over the plant parts as sweet material which was liked by the ants for feeding and due to the association of ants the above homopterous insects were escaped from the natural enemy attack.

Table 1: Abundance of Ants from Western Maharashtra (Kolhapur, Sangli and Satara)

Sl.No.	*Species*	*Abundance*	*Locality*	
			Agro-ecosystem	*Forest Ecosystem*
	Family – Formicidae			
	Sub-family – Formicinae			
1.	*Anoplolepis gracilipes* Smith	Common	K, Sn, St	–
2.	*Paratrechina ankarana* Lapolla and fisher	Common	K, Sn, St	K, Sn, St
3.	*Oecophylla smaragdina* Fabricius	Common	K, Sn, St	K, Sn, St
4.	*Camponotus compressus* Fabricius	Common	K, Sn, St	–
5.	*Camponotus irritans* Smith	Common	K, Sn, St	-
6.	*Camponotus variegates* Smith	Common	–	K, Sn, St
7.	*Camponotus sericeus* Fabricius	Common	–	K, Sn, St
8.	*Camponotus cinerascens* Fabricius	Common	–	K,Sn,St
9.	*Camponotus maculates basalis* Smith	Rare	–	K,Sn,St
10	*Camponotus infuscus* Forel	Common	–	K,Sn,St
11.	*Camponotus maculates pallidus* Smith	Common	–	K,St
12.	*Camponotus radiates* Forel	Rare	–	K, Sn, St
13.	*Camponotus selene* Emery	Rare	–	K, Sn, St
14.	*Camponotus variegates kattensis* Bingham	Rare	–	K, Sn, St
15.	*Polyrachis dives* Smith	Common	K, Sn, St	K, Sn, St
16.	*Polyrachis convexa* Roger	Common	K, Sn, St	K, Sn, St
17.	*Polyrachis* sp.	Rare	K, Sn, St	–
18.	*Formica gravelyi* Mukerjee	Common	K, Sn, St	K, Sn, St
	Sub-family- Ponerinae			
19.	*Harpegnathos saltator* Jerdon	Rare	–	K, Sn, St
20.	*Leptogenys chinensis* Mayr	Rare	–	K, Sn, St
21.	*Dinoponera australis* Emery	Rare	–	K, Sn, St

Contd...

Table 1–*Contd...*

Sl.No.	Species	Abundance	Locality	
			Agro-ecosystem	Forest Ecosystem
	Sub-family – Dorylinae			
22.	*Dorylus labiatus* Shuckard	Common	K, Sn, St	K, Sn, St
23.	*Dorylus orientalis* Westwood	Common	K, Sn, St	K, Sn, St
	Sub-family – Dolichoderinae			
24.	*Tapinoma melanocephalum* Fabricius	Common	K, Sn, St	K, Sn, St
25.	*Dolichoderus affinis* Emery	Common	K, Sn, St	K, Sn, St
26.	*Dolichoderus sundari* Mathew and Tiwari	Common	K, Sn, St	K, Sn, St
	Sub-family – Pseudomyrmecinae			
27.	*Tetraponera rufonigra* Jerdon	Common	K, Sn, St	K, Sn, St
	Sub-family – Myrmicinae			
28.	*Cataulacus taprobanae* Smith	Common	K, Sn, St	K, Sn, St
29.	*Crematogaster rogenhoferi* Mayr	Common	K, Sn, St	K, Sn, St
30.	*Crematogaster ashmeadi* Mayr	Common	K, Sn, St	K, Sn, St
31.	*Crematogaster abdominalis* Motschoulsky	Common	K, Sn, St	–
32.	*Crematogaster anthracina* Smith	Common	–	K, Sn, St
33.	*Crematogaster diffusa* Jerdon	Common	–	K, Sn, St
34.	*Crematogaster himalayana* Forel	Common	–	K, Sn, St
35.	*Crematogaster politula* Forel	Common	–	K, Sn, St
36.	*Crematogaster pradipi* Tiwari	Common	–	K, Sn, St
37.	*Myrmecaria* sp.	Common	K, Sn, St	K, Sn, St
38.	*Myrmecaria* spp.	Common	K,Sn,St	K,Sn,St
39.	*Solenopsis geminata* Fabricius	Common	K, Sn, St	K, Sn, St
40.	*Monomorium minutum* Buckely	Common	K, Sn, St	K, Sn, St
41.	*Monomorium destructor* Jerdon	Common	–	K, Sn, St
42.	*Monomorium glabrum Andr.*	Common	K, Sn, St	–
43.	*Monomorium indicum Forel*	Common	K, Sn, St	–
44.	*Monomorium gracillinum*	Common	K, Sn, St	–
45.	*Myrmica beesoni* Mukerjee	Common	–	K, Sn, St
46.	*Myrmica indica* Weber	Common	K, Sn, St	–
47.	*Myrmica rugifrons* Smith	Common	–	K, Sn, St
48.	*Myrmica rugosa* Mayr	Common	K, Sn, St	–
49.	*Myrmica smythiesii* Forel	Common	–	K, Sn, St
50.	*Pheidole megacephala* Fabricius	Common	K, Sn, St	K, Sn, St
51.	*Pheidole ghatica* Forel	Common	–	K, Sn, St
52.	*Pheidole indica* Mayr	Common	K, Sn, St	–
53.	*Pheidole indica coonoorensis* Forel	Common	–	K, Sn, St
54.	*Pheidole indica rotschana* Forel	Common	–	K, Sn, St
55.	*Aphaenogaster beccari* Emery	Rare	-	K, Sn, St

K: Kolhapur; Sn: Sangli; St: Satara.

Plate 32. **Figure 142: Nest of *Solenopsis* sp.; Figure 143: Nest of *Monomorium* sp.; Figure 144: Nest Door of *Crematogaster* sp.; Figure 145: *Polyrachis* sp. Nesting on *M. acumilata*; Figure 146: Nesting door of *Aphaenogaster* sp.**

Plate 33. Figure 147: Nest of *C. taprobanae* on *M. indica;* Figure 148: *O. smaragdina* Nest on *S. asoca*; Figure 149: Nest of *Crematogaster* sp. on *M. indica*; Figure 150: Nest of *Camponotus* sp.; Figure 151: Nest of *Camponotus* sp. in Jawar ecosystem; Figure 152: Author collecting insects.

Plate 34. Figure 153: ***Camponotus*** **sp. on** ***A. esculentus;*** **Figure 154:** ***Crematogaster*** **sp. on** ***F. racemosa;*** **Figure 155:** ***Camponotus*** **sp. on** ***Z. mays;*** **Figure 156:** ***Camponotus*** **sp. on** ***S. bicolor;*** **Figure 157:** ***Camponotus*** **sp. on** ***C. cajan.***

There were two types of ants, winged and non-winged. The winged ants have sting poison gland opening into the sting while non-winged have poisonous secretions in sharp mandibles. In both cases formic acid was responsible for irritation and itching and swelling on the human body. Non poisonous ants were associated with sweet materials in human houses. They contaminated food and were responsible for transmition of certain diseases to humans.

Discussion

A total of 36 species of ants were recorded by Bhoje *et al.* (2014) from Kolhapur district out of which the members of genera *Monomorium, Dolichoderus, Formica* and *Camponotus* were found in human habitat. *Formica* species were associated with some vegetable crops like brinjal and chilli. The species *Crematogaster himalayana, C. anthracina, Phiedole ghatica* and *P. indica* were found in forests of Western Ghats. Most of the species were recorded from forest regions of Kolhapur districts.

Seasonal patterns of ants were analyzed by Bharti *et al.* (2009) from Punjab Shivalik range of North Western Himalaya. Various collection methods like pitfall traps, Winkler's fish bait and hand picking were used. They observed 40 species belonging to 8 sub-families as seasonal pattern. The sub-family Myrmicinae followed by Formicinae were found to be dominant. Further, they noted that temperature and relative humidity were correlated with seasonal distributional patterns.

A seasonal variation in ant activity was studied by Kharbani and Hajong (2013) from a subtropical humid forest of Meghalaya, northeast India, using pitfall traps

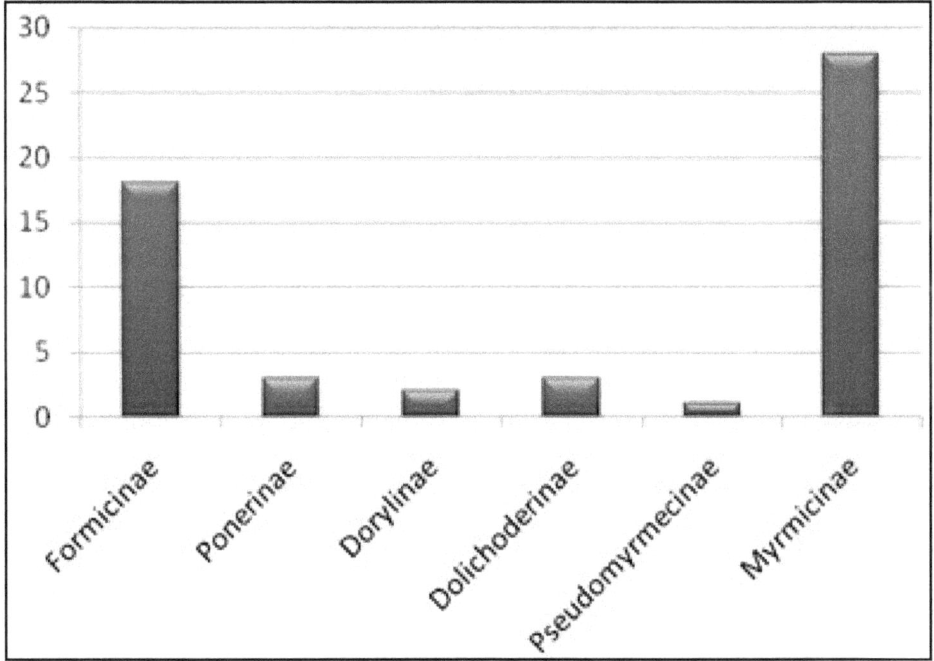

Figure 158: Seasonal Abundance of Ants from Western Maharashtra.

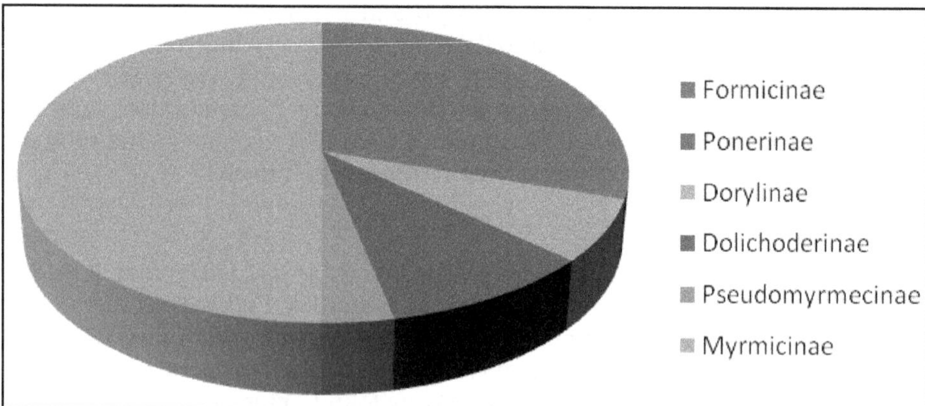

Figure 159: Seasonal Abundance of Ants in Agro-ecosystem.

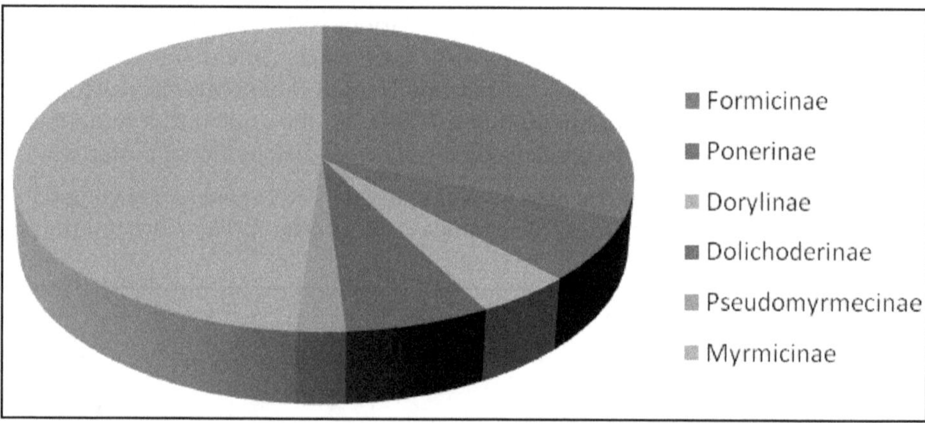

Figure 160: Seasonal Abundance of Ants from Forest Ecosystem.

as the sampling method. A total of 108,733 ant individuals were collected during sampling, consisting 28 species and 18 genera. Sub-families comprised Myrmicinae (12 spp.), Formicinae (8 spp.), Ponerinae (5 spp.), Dorylinae, Cerapachyinae and Dolichoderinae (1 sp each). Number of species recorded was highest during spring and summer and lowest during winter. Number of individuals and species diversity per sample was lower during winter and higher during summer. Number of individuals, species richness and Shannon entropy per sample varied significantly among seasons. In the present study, the sub-family Myrmicinae was dominant over others and the sub family Formicinae ranked second as far as seasonal abundance is concerned. Kharbani and Hajong (2013) further reported that seasonal changes in ant individual numbers, species richness and Shannon- Wiener index was significantly positively correlated with soil temperature, relative humidity, air temperature and soil moisture, but not significantly correlated with soil pH.

A survey of ant faunal diversity, species composition and the effect of microhabitat were studied by Ramesh *et al.* (2010) from the Department of Atomic Energy(DAE) Campus, Kalpakkam, South India during dry season (March-June 2008). Ants were collected by pitfall trap and hand picking in five habitats with different vegetation type. Totally 35 species of Formicidae belonging to 5 sub-families were reported. The Myrmicinae was dominant over other by species diversity (18 spp) and abundance. The species rich genera were *Monomorium, Camponotus, Tetraponera, Crematogaster* and *Tetramorium.* Their results indicated that the scrub jungle and riparian woods were most diverse habitats followed by monoculture and sandy area. Vegetation type, soil characteristics and anthropogenic disturbances were chief factors for influence on ant assemblage.

Kumar and Mishra (2008) studied ant community variation in urban and agricultural ecosystems in Vadodara districts of Gujarat State, India. They used pitfall traps and hand collection methods for ants. A total of 22 ant species from 13 genera and 5 sub-families were collected. Ants of 9 genera and 15 species were found in urban ecosystems and 10 genera and 16 species in agricultural ecosystems. The genus *Pheidole* showed greatest species richness in urban ecosystems, where as *Camponotus* and *Pheidole* were equally specious in the agricultural ecosystems. The composition of species was unique to each habitat and mostly governed by the vegetation and associated biota.

Nagariya and Pawar (2014) studied the distribution of ants diversity in Pohara forest area of Amravati, Maharashtra, India. They reported 3 species and 3 genera of ants from the family Formicidae. The dominant species reported by them refer to red imported fire ant (*Solenopsis invicta*), Carpenter ant (*Camponotus* sp.) and Pharaoh ant (*Monomorium pharaonis*).

Bhoje *et al.* (2014) studied the diversity of ants from Amba reserve forest of Western Ghats. They reported 65 species of ants belonging to 35 genera of 8 sub-families. Again, the sub-family Myrmicinae was dominant over other sub-families in Amba Ghats.

Recently, Kurane *et al.* (2015) studied the diversity and economic importance of ants (Hymenoptera: Formicidae) from Kolhapur city, Maharashtra State, India. With their economic importance, they reported 20 species of ants belonging to the genera *Camponotus, Monomorium, Crematogaster, Dolichoderus, Formica* and *Dorylus.*

Seasonal assemblage of leaf litter ants in Megamalai, Western Ghats, India was studied by Selvarani *et al.* (2014). They reported that the summer season was abundant by member of sub-family Ponerinae and Myrmicinae with respect to *Diacamma rugosum* and *Monomorium glabrum*. Similarly, in autumn season *i.e.* post monsoon season, high abundance of *Lepisiota* sp., *Diacamma rugosum*, members of Formicinae and Ponerinae sub-family respectively were noticed. While, Khot *et al.* (2013) studied the ant diversity from an urban garden of Mumbai, Maharashtra. They noticed that 28 species of ants representing 6 sub-families namely, Aenictinae, Dolichoderinae, Formicinae, Myrmicinae, Ponerinae and Pseudomyrmicinae were prevalent in the gardens. The highest diversity was exhibited by the sub-family Myrmicinae with 11 ant species and 7 genera.

In the present study, a total 55 species belonging to 21 genera and 6 sub-families were reported and the sub-family Myrmicinae was dominant over others. Since ants are very economically important group, the present work will add great relevance in protecting and conserving and utilizing the biodiversity of Western Maharashtra.

6

Biology, Ecology and Control of Ants

Introduction

Ants (Hymenoptera: Fornicidae) are social insects and found in all terrestrial habitats. Their morphology and size varied as per the habitat. They are docile to extremely pugnacious. Ants feed on plant seeds, nector, honey drew secreted by cell sap sucking insects and fungi. They may be carnivores, scavenger, nacrophages, commonsalizers, predators and pests on economically importance crop plants. They also plays an important role in food web and have been studied by Agosti (1991), Arnold (19115), Bingham (1903), Bolton and Collingwood (1975), Wu (1990) etc. Ants play key ecological functions in nature. They act as biocontrol agents in many insect pests in agriculture, horticulture and forest ecosystems. They are used for biodiversity assessment and comparison of habitats of ecosystems. Since, ants are social insects, their life cycle pattern has also tremendous importance in their control in various crop ecosystems. The black ants are commonest troublesome house hold pests in India and different parts of the world.

Following ant species are commonly found in India which invades houses and godowns.

1. The small red house ant and *Dorylus labiatus* Shuekard (small and red)
2. *Monomorium indicum* F. (small and brown)
3. *M. aracollium* (small and brown)
4. The carpenter ant *Camponotus compressus* Fabricius (large and black)

Ants are social insects. They live in colonies, each colony consist 3 casts namely king, queen and workers. There is polymorphism, division of labour and swarming in ants.

Swarming

Swarming is emergence of winged ants with large number which is aimed for mating and establishing new colony. Swarming is occurred in monsoon. The winged males and female as young queens recognised the pair and mate in the air. After mating young female (queens) descend down on the soil and shed their wings and start constructing small nest. By the time, the males die or may get survival. However, the young queen prepared cavity like nest and after completion, it sealed the door and remain inside for sometime after waiting for maturation of eggs in its body, then eggs are laid. Thus, pre oviposition lasts for one or few months. She lay eggs inside the nest (chamber).

Life-Cycle

The general pattern of life cycle of ants shows 4 distinct stages *viz.* egg, larva (grub), pupa and adult.

Eggs

Eggs are typically hymenopteriform, whitish and are more elongated and slightly tapers and rounded at both the ends (Figure 158), eggs hatch in about one week.

Larvae (Grub) (Figure 159)

The larvae or grubs are whitish or opaque in colour without legs. There are five instrars in the larval stage. Queen feeds her salivary secretion to grubs. Such secretions have her own body fats and wing muscles with certain enzymes. Larval period is approximately 10 to 20 days depending on climatic conditions. Full grown larvae pupate as exarate pupae (Figure 160). Initially the pupa is whitish and there are three distinct divisions of body parts *viz.*, head, thorax and abdomen with developing wing pads, legs and antennae (Figure 160). Later, the pupa becomes yellow/brown/black at the time of hatching into adult. Pupal period is 8-12 days depending or climatic conditions and first batch of adult workers is emerged. Such workers are sterile females and are involved in the duties as per the division of labour. On priority basis, after emergence as an adults, they supply the food to the colony. However, the first progeny is governed by the queen.

Adults

There is polymorphism in ants. However, in ants, adults are characterized by two pairs of transparent wings, petiolate abdomen and geniculate antennae. Wingless and stingless adult ants are also found.

Since workers are involved in providing food to the developing individuals and royal pair, subsequent generation individuals emerged are large sized and healthy and responsible for increasing the size of colony. King (males) and queen are only involved in reproductive activities in the colony.

The nests of ants are always found open outside much away from the queen chamber. However, in termites, although they are social insects, their nests never open outside, infact, they are closed. The life of colony and the queen is considerably longer. The queen can survive for 15 years.

Ecology of Ants

Ants play a major role in most terrestrial ecosystems by performing key ecological functions. They act as predators of insects, scavengers, carnivores, commonsalizers, nacrophages, insect pest control agents, food and medicine for humans. Therefore, ants are supposed to be the most important group of insects for their study of conservation, protection and utilization in the development of a region or a country.

The ants are found in both agro and forest ecosystems; their diversity study was illustrated comparatively even in urban and agricultural ecosystems (Kumar and Mishra, 2008). In Western Maharashtra *C. compressus* and *Phiedole* spp. always constructed their nests around the roots of *Tamarindus indicus. C. compressus* also constructed its nests with the association of *Acacia nilotica* and *Moringa oleifera. C. compressus* was also associated with *Ficus carica* (Angeer), *Mangifera indica* (Mango), Jamun, Guava, China rose *etc.* The above species of ants and many others represented in text were found with association of Jassids, Scales, White flies, Mealy bugs, Psyllids, Catterlpillars and Grasshoppers *etc.* as a part of commensalization or predation.

Tetraponera species of ants have foraged and nested on trees like *Caesalpinia crista.* The *Monomorium* spp. and *Solenopsis* spp. have foraged inside the fields and nested at field margins while; *Crematogaster* spp., *Polyrachis* sp. and *Leptogenys* sp. were found foraging on large trees like *C. crista* as solitary individuals. *O. smaragdina* was associated with Gulmohar tree in Kolhapur region and feeding on Lepidopteraous caterpillars while, *Formica* spp. and *Camponotus* spp. were in association of aphids on many economically important plants like Mango, Angeer, Guava, Jamun, Banana, Chilli etc. The small sized insects like *Monomorium* spp. and *Solenopsis* spp. were in association with aphids, scales, mealy bugs and psyllids on brinjal, anger, mango and guava. However, more investigations are needed on pest, predatory and commensalization of ants which will be helpful for protecting diversity and biodiversity byproducts produced of the region.

Nature of Damage

Out of three casts found in the colony of ants, workers are the real destructives creatures. They collect and destroy every thing which is edible. However, they like sweet materials and fats. But, they are equally competent to other grain pests for destroying grains and seeds. The ants are responsible for leakage of human houses during monsoon since they make holes to the walls, basement and roofs of human dwellings. They also makes holes to wooden materials used in building houses and household goods. In recent years, they are visualized as pets of several agricultural crops like brinjal, chili, etc. They girdle bark of the crop from ground level, prepare their nesting at the base of crop plants and affect the growth and yield of the crops.

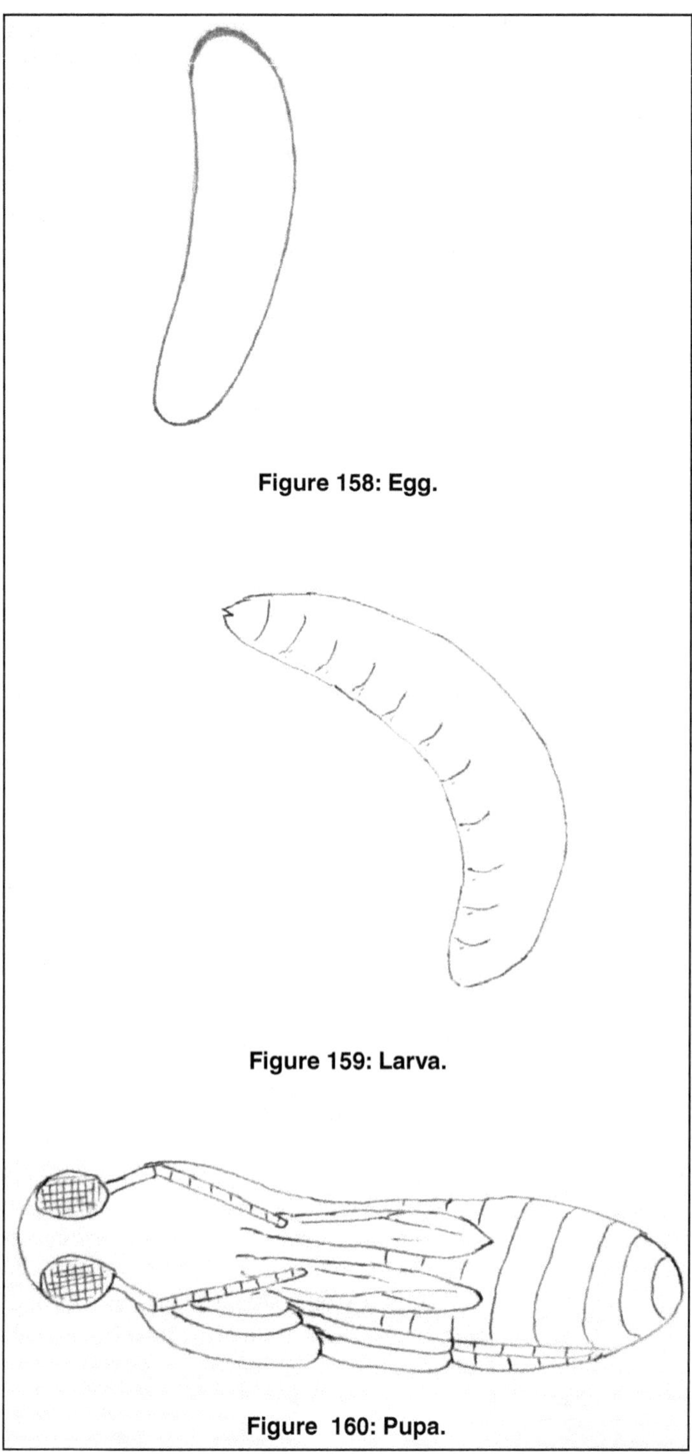

Figure 158: Egg.

Figure 159: Larva.

Figure 160: Pupa.

Ant Life Stages.

Control Strategies for Ants

1. The colony of ants should be destroyed by digging the soil.
2. The queen and king should be identified and destroyed with the help of insect net or pestisides like 10 per cent BHC emulsion or Chlordane 5 per cent spray on nesting.
3. On edible crops Melathion 0.5 – 1 per cent spray be used.
4. In human dwelling and store houses Ant chokes are used by making border by choks to the infested area by keeping ants in prison for death.
5. For agricultural crops irrigation, specially sprinkling method can suppress the pest population with out causing pollution in agro ecosystems.

7

Summary and Conclusion

Ants belong to Order Hymenoptera, containing about 9000 described species from world. They are social insects and found in all terrestrial habitats. Ants are visualized as pests of agricultural, horticultural and forest plants and used in biological pest control of insect pests. Ants have nutritional and medical importance. Hence knowledge on their diversity, seasonal abundance and distribution and host records will add great relevance for understanding ants for its management and their utility in biological pest control and protecting biodiversity of Western Maharashtra. Hence the present topic was selected.

This book has been divided into 5 chapters. First chapter is devoted for introduction which embodies characteristics of agro and forest ecosystems (Western Ghats) and importance of ants. It also includes national and international status of the topic. Second chapter is devoted for the review of literature on the topic.

The chapter third deals with materials and methods adapted for completion of work. The ants have been collected with the help of hand picking method. Collected specimens have been preserved in insect storage box and identified by consulting appropriate literature. Abundance and distribution of ants has been made by visiting and collecting ants from the study spots at 15 days interval by one man one hour search method. Morphological biodiversity has been studied by studying body Parts of ants such as head, thorax, abdomen and their appendages.

Fourth chapter embodies the diversity of ants which includes morphological description of 21 species out of which three were found new to the science. The species described in the chapter refers to *Anoplolepis gracilipes* Smith, *Paratrechina ankarana* Lapolla and fisher, *Oecophylla smaragdina* Fabricius, *Camponotus compressus* Fabricius, *Camponotus irritans* Smith, *Camponotus variegates* Smith, *Polyrachis dives* Smith, *Polyrachis convexa* Roger, *Harpegnathos saltator* Jerdon, *Leptogenys chinensis* Mayr, *Tapinoma melanocephalum* Fabricius, *Tetraponera rufonigra* Jerdon,

Crematogaster ashmeadi Mayr, *Crematogaster rogenhoferi* Mayr, *Cataulacus taprobanae* Smith, *Solenopsis geminata* Fabricius, *Monomorium minutum* Buckely, *Pheidole megacephala* Fabricius and *Aphaenogaster beccari* Emery and three new species refer to the genera *Polyrachis* and *Myrmecaria*, especially *Polyrachis indica* sp.n., *Myrmicaria sinensis* sp.n. and *Myrmicaria kolhapuriensis* sp.n.

The chapter fifth is related to seasonal abundance of the ants. A total of 55 species belonging to 21 genera and 6 sub-families were reported and the sub-family Myrmicinae was dominant over other five namely, Formicinae followed by Ponerinae followed by Dolichoderinae, Dorylinae and Pseudomyrmecinae. Since ants are very economically important group, the present work will add great relevance in protecting and conserving and utilizing the biodiversity of Western Maharashtra. The chapter sixth contains biology, ecology and control of ants. Seventh chapter contains summery and conclusion of the text. Lastly, the text contains bibliography referred on the topic.

Conclusion

Ants are social insects and found in all terrestrial habitats. They are visualized as pests of agricultural, horticultural and forest plants and used in biological pest control of insect pests. Ants have nutritional and medical importance. Hence knowledge on their diversity, seasonal abundance and distribution and host records will add great relevance for understanding ants for its management and their utility in biological pest control and biodiversity conservation and protection.

Bibliography

Agosti, D, 1994. The Phylogeny of the ant tribe *Formicini* (Hymenoptera: Formicidae) with the description of a new genus. *Systematic Entomology,* 19: 93-117.

Agosti, D. 1991. Revision of the Oriental ant genus *Cladomyrma,* with an outline of the higher classification of the Formicinae. *Systematic Entomology,* 16: 293-310.

Agosti, D. 1995. A revision of the South American species of the ant genus *Probolomyrmex. Journal of the New York Entomological Society,* 102 : 429-434.

Agosti, D. and B. Bolton, 1990. New Characters to differentiate the ant genera *Lasius F.* and *Formica L.* (Hymenoptera: Formicidae). *Entomologist's Gazettee,* 41.

Agosti, D., Majer, J.D., Alonso, A.E. and T.R. Schultz (Eds), 2000. Ants: Standard methods for measuring and monitoring biodiversity, Smith Sonian Institution.

Agosti, D., Maryati, M. and CYC, Arthur, 1994. Has the diversity of tropical ant fauna been underestimated? An indication from leaf litter studies in a West Malaysian Lowland rain forest. *Tropical Biodiversity* 2:270-275.

Agosti, D.and Johnson, N F, 2003. *La nueva taxonomía de hormigas. 45–48 in Fernández, F. Introducción a las hormigas de la región neotropical.* Instituto Humboldt, Bogotá.

Aktaç, N. 1977. Studies on the myrmeco fauna of Turkey. 1. Ants of Siirt, Bodrum, and Trabzon. Istanbul Universitesi Fen Fakültesi Mecmuasi. Seri B. 41 (1976): 115-135. *Cladomyrma,* American Museum Novitates 3283: 1-24.

Ali, T.M, 1991. Ant fauna of Karnataka. *IUSSI Newsletter.*5:1-8.

Ali, T.M, 1992. Ant fauna of Karnataka. *IUSSI Newsletter.*6:1-7.

Alpert, G.D. 1992. Observations on the genus *Ërataner* in Madagascar. *Psyche,* 99: 117-127.

Andersem,A.N and R.E.Clay,(1996). The ant fauna of Danggali Conservation Park insemi-arid South Australia: a comparison with Wyperfield(Vic.) and Cape Arid (W.A.) National Parks. *Aust. J. Entom.* 35,289-295.

Andersen, A.N.(1982). Seed removal by ants in the malle of north western Victoria. In *Ant-Plant Interactions in Australia.* (R.C.Buckley,ed).pp.31-43.The Haughe, Netherlands: Dr. Junk Publishers.

Anderson, A.N. and R.E. Clay, 1996. The ant fauna of Danggali Conservation Park in semiarid South Australia: A comparison with Wyperfield (Vic) and Cape arid (W.A.) National Parks *Aust. J. of Entomology,* 35: 289-295.

Anderson, K.E., Russell, J.A., Moreau, C.S.,Kautz, S., Sullam, K.E., Hu, Y., Basinger, U., Mott, B.M., Buck, N and D.E. Wheeler, 2012. "Highly similar microbial communities are shared among related and trophically similar ant species". *Molecular Ecology* **21**: 2282–2296. doi:10.1111/j.1365-294x.2011.05464.x.

Anto, A. and Sabu, Thomas K, 2007. Biodiversity analysis of Forest litter ant assemblages in the Wayanad region of Western Ghats using taxonomic and conventional diversity measures. *Journal of Insect Science,* 7(6):1-13.

Arnold, G. 1915. A monograph of the Formicidae of South Africa. Part 1. (Ponerinae; Dorylinae.) *Annals of the South African Museum,* 14: 1-159.

Arnold, G. 1916. A monograph of the Formicidae of South Africa. Part 2. (Ponerinae; Dorylinae.) *Annals of the South African Museum,* 14: 159-270.

Arnold, G. 1917. A monograph of the Formicidae of South Africa. Part 3. (Myrmicinae.) *Annals of the South African Museum,* 14: 271-402.

Arnold, G. 1920a. A monograph of the Formicidae of South Africa. Part 4. (Myrmicinae.) *Annals of the South African Museum,* 14: 403-578.

Arnold, G. 1920b. In Santschi, F. Formicides africains et américains nouveaux. *Annales de la Société Entomologique de France,* 88: 361-390.

Arnold, G. 1922. A monograph of the Formicidae of South Africa. Part 5. (Myrmicinae.) *Annals of the South African Museum,* 14: 579-674.

Arnold, G. 1924. A monograph of the Formicidae of South Africa. Part 6. (Camponotinae.) *Annals of the South African Museum,* 14: 675-766.

Arnold, G. 1926. A monograph of the Formicidae of South Africa. Appendix. *Annals of the South African Museum,* 23: 191-295.

Ashmead, W.H. 1905a. New Hymenoptera from the Philippines. *Proceedings of the United States National Museum,* 29: 107-119.

Ashmead, W.H. 1905b. A skeleton of a new arrangement of the families, sub-families, tribes and genera of the ants, or the superfamily Formicoidea. *Canadian Entomologist,*37: 381-384.

Ashmead, W.H. 1906. Classification of the foraging and driver ants, or family Dorylidae, with a description of the genus *Ctenopyga* Ashm. *Proceedings of the Entomological Society of Washington,*8: 21-31.

Astruc, C., Julien, J.F., Errard, C, and Lenoir, A. 2004. Phylogeny of ants based on morphology and DNA sequence data. *Molecular Phylogenetics and Evolution* 31: 880-893.

Basu, P, 1997. Seasonal and spatial pattern in ground dwelling ants in a rain forest in the Western Ghats, India. *Biotropica*, 29(4):489-500.

Beason, C.F, 1941. Ecology and Control of forest insect of India and neighbouring countries, Govt. of India (1961 Reprint), pp.767.

Beatson, S.H., 1972. Pharaoh's ants pathogen vectors in hospitais. *The Lancet*; 1:425-427.

Bequaert, J,1921. "Insects as food: How they have augmented the food supply of mankind in early and recent times". *Natural History Journal* **21**: 191–200.

Berghoff, S.M. and N.R. Franks, 2007. First record of the army ant *Cheliomyrmex morosus* in Panama and its high associate diversity. *Biotropica,* 10(1111). 1744-7429.

Bewa, K., Das, A., Krishnaswamy, J., Karanth, K.U., Kumar S.N., Rao,M., Bhargav,P., Ganeshaiah,K.N and V Srinivas, 2007. Western Ghat and Sri Lanka Biodiversity Hotspots: Western Ghat Region. Critical Ecosystems Partnership Fund, May.

Bharti, H, 2011.List of Indian ants.*Halters*, 2:79-87.

Bharti, H. and Kumar, R. 2012a. *Lophomyrmex terraceensis*, a new ant species of the bedoti group with a revised key. *Journal of Asia-Pacific Entomology*, 15: 265-267.

Bharti, H. and Kumar, R. 2012b. Taxonomic studies on genus *Tetramorium* Mayr, with report of two new species and three new records including a tramp species from India with a revised key. *ZooKeys*, 207: 11-35.

Bharti, H. and Kumar, R. 2012c. A new species of *Leptanilla* with a key to Oriental species. *Annales Zoologici* (Warszawa), 62: 619-625.

Bharti, H. and Wachkoo, A.A. 2011. *Amblyopone boltoni,* a new ant species from India. *Sociobiology*,58: 585-591.

Bharti, H. 2001. Two new species of *Pheidole* Westwood from India. *Journal of Entomological Research* (New Delhi), 25: 243-247.

Bharti, H. 2003a. Queen of the army ant *Aenictus pachycerus*. *Sociobiology*,42: 715-718.

Bharti, H. 2003b. *Polyrhachis punjabi* sp. n. from India. *Folia Heydrovskyana*, 11 (1): 1-3.

Bharti, H. 2003c. A new species of Crematogaster from India. *Entomon*, 28: 85-88.

Bharti, H. 2008. Redescription of *Crematogaster subnuda subnuda* Mayr, 1879. *Journal of Entomological Research*,32: 83-88.

Bharti, H. 2012b. Two new species of the genus *Myrmica* from the Himalaya. *Tijdschrift voor Entomologie,* 155: 9-14.

Bharti, H. and A. A. Wachkoo, 2014a. Anew carpenter ant, *Camponotus parabarbatus* (Hymenoptera: Formicidae) from India. *Biodiversity Data Journal,* 2:e996. doi:10.3897/*BDJ*.2e996.9 pp.

Bharti, H. and Akbar, S.A, 2014. *Meranoplus periyarensis*, a remarkable new ant species (Hymenoptera: Formicidae) from India. *Journal of Asia – Pacific Entomology*, 17:811-815.

Bharti, H. and S.A. Akbar, 2013. Taxonomic studies on the ant genus *Cerapachys* Smith (Hymenoptera: Formicidae) from India. Zookeys, 336:79-103.doi:10.3897/zookeys.336.5719.

Bharti, H. and S.A. Akbar, 2014. *Meranoplus periyarensis*, a remarkable new ant species (Hymenoptera: Formicidae) from India. *Journal of Asia - Pacific Entomology*, 17: 811-815.

Bharti, H., Gul, I. and Schulz, A. 2012. Three new species of genus *Temnothorax* from Indian Himalayas with a revised key to the Indian species. *Acta Zoologica Academiae Scientiarum Hungaricae*,58: 325-336.

Bharti, H., Sharma, Y. P. and A.Kaur, 2009. Seasonal patterns of ants (Hymenoptera: Formicidae) in Punjab Shivalik. *Halters*, 1(1): 36-47.

Bharti, H., Wachkoo, A. A. and R. Kumar, 2012. Two remarkable new species of *Aenictus* (Hymenoptera: Formicidae) from India. *Journal of Asia-Pacific Entomology*, *15*: 291-294.

Bharti, H. 2012a. Myrmica *H. nefaria* sp. n. – a new social parasite from Himalaya. *Myrmecological News*, 16: 149-156.

Bhoje, P.M., Kurane, S.H., Desai, A.S and T.V.Sathe, 2014. Biodiversity of Ants (Hymenoptera: Formicidae) of Amba Reserve Forest of Western Ghats, Maharashtra. *Global Journal for Research Analysis*, 3(7):1-4.

Bhoje, P.M, Kurane, S.H. and T.V.Sathe, 2014. Diversity of Ants (Hymenoptera: Formicidae) From Kolhapur District of Maharashtra, India. *Uttar Pradesh Journal of Zoology*, 34(1): 23-25.

Bingham, C.T. 1903a. The Fauna of British India, Including Ceylon and Burma (Hymenoptera: Formicidae), 2: 1-414.

Bingham, C.T. 1903b. *Fauna of British India. Hymenoptera, 3*, pp- 311.

Biodiversity of Ants in Bangalore. *J. Ecobio:* 12(2): 115-122.

Bollazzi M., Kronenbitter J., Roces F. (2008) Soil temperature, digging behaviour, and the adaptive value of nest depth in South American species of *Acromyrmex* leaf-cutting ants, *Oecologia* 158, 165–175

Bolton, B, 1977. The ant tribe Tetramoriini (Hymenoptera: Formicidae). The genus *Tetramorium* Mayr in the Oriental and Indo-Australian regions and Australia. *Bull. Br. Mus. Nat. Hist. (Ent.)*, 36 (2): 67-151, illustr.

Bolton, B, 1978. The ant tribe Tetramoriini (Hymenoptera: Formicidae). Constituent genera, review of smaller genera and revision of *Triglyphothrix* Forel. *Bull Br. Mus. Nat. Hist. (Ent.)*, 34 (5) 1976: 283-379.

Bolton, B, 1980. The ant tribe Tetramoriini (Hymenoptera: Formicidae). The genus *Tetramorium* Mayr in the Ethiopian Zoogeographical region. *Bull. Br. Mus. Nat. Hist. (Ent)*, 40 (3): 193-384, illustr.

Bolton, B, 1982. Afrotropical species of the Myrmicinae ant genera *Cardiocondyla, Leptothorax, Melissotarsus, Messor* and *Cataulacus*. *Bulletin of the British Museum (Natural History)(Entomology)*, 45: 307-37.

Bolton, B, 1987. A review of the *Solenopsis* genus-group and revision of Afro-tropical *Monomorium* Mayr (Hymenoptera: Formicidae). *Bull. Br. Mus. Nat. Hist. (Ent.)*, 54 (3) : 263-264 illustr. (In English).

Bolton, B. 1988. A review of Pamtopula Wheeler, a forgotten genus of Myrmicine ants (Hymenoptera: Formicidae). *Entomol. Man. Mag.*, 124: 125-143, illustr. (In English). 94 *Memoirs of the Zoological Survey of India.*

Bolton, B, 1994. *Identification Guide to the Ant Genera of the World.* Havard University Press, Cambridge, Massachusetts, USA.

Bolton, B, 1994. Identification Guide to the Ant Genera of the World. Harvard University Press, Cambridge, Massachusetts, USA.

Bolton, B, 1995. A taxonomic and Zoogeographical census of the extant and taxa (Hymenoptera: Formicidae) *J. Nat. Hist.*, 29: 1037-1056.

Bolton, B. and Belshaw, R. 1993. Taxonomy and biology of the supposedly lestobiotic ant genus *Paedalgus. Systematic Entomology*,18: 181-189.

Bolton, B. and Brown, W.L.,Jr. 2002. *Loboponera* gen. n. and a review of the Afrotropical *Plectroctena* genus group. *Bulletin of The Natural History Museum (Entomology Series)*, 71: 1-18.

Bolton, B. and Collingwood, C.A. 1975. Hymenoptera: Formicidae. Handbooks for the Identification of British Insects 6, part 3 (c): 34 pp. London.

Bolton, B. and Fisher, B.L. 2008a. The Afrotropical Ponerinae ant genus *Asphinctopone* Santschi. *Zootaxa*, 1827: 53-61.

Bolton, B. and Fisher, B.L. 2008b. The Afrotropical Ponerinae ant genus *Phrynoponera* Wheeler. *Zootaxa*, 1892: 35-52.

Bolton, B. and Fisher, B.L. 2008c. Afrotropical ants of the Ponerinae genera *Centromyrmex* Mayr, Promyopias Santschi gen. rev. and Feroponera gen. n., with a revised key to genera of African Ponerinae. *Zootaxa*,1929: 1-37.

Bolton, B. 2011. Bolton's Catalogue and Synopsis, inhttp://gap.entclub.org/Version: 1 July, 2011.

Bolton, B. 2013. An online catalog of the Ants of the World. *Antcat.* http://www.antcat.org/(accessed July 14, 2013).

Bolton, B. 1971. Two new subarboreal species of the ant genus *Strumigenys* from West Africa. *Entomologist's Monthly Magazine*, 107: 59-64.

Bolton, B. 1972. Two new species of the ant genus *Epitritus* from Ghana, with a key to the world species. *Entomologist's Monthly Magazine*,107 (1971): 205-208.

Bolton, B. 1973a. The ant genera of West Africa: a synonymic synopsis with keys. *Bulletin of the British Museum (Natural History) (Entomology)*, 27: 317-368.

Bolton, B. 1973b. The ant genus *Polyrhachis* F. Smith in the Ethiopian region. *Bulletin of the British Museum* (*Natural History*) (*Entomology*),: 283-369.

Bolton, B. 1973c. A remarkable new arboreal ant genus from West Africa. *Entomologist's Monthly Magazine*,108 (1972): 234-237.

Bolton, B. 1974a. A revision of the Palaeotropical arboreal ant genus *Cataulacus* F. Smith. *Bulletin of the British Museum* (*Natural History*) (*Entomology*), 30: 1-105.

Bolton, B. 1974b. New synonymy and a new name in the ant genus *Polyrhachis* F. Smith. *Entomologist's Monthly Magazine*, 109: 172-180.

Bolton, B. 1974c. A revision of the ponerine ant genus *Plectroctena* F. Smith. *Bulletin of the British Museum* (*Natural History*) (*Entomology*), 30: 309-338.

Bolton, B. 1975a. A revision of the ant genus Leptogenys Roger in the Ethiopian region, with a review of the Malagasy species. *Bulletin of the British Museum (Natural History) (Entomology)* 31: 235-305. [5.ii.1975.]

Bolton, B. 1975b. A revision of the African ponerine ant genus Psalidomyrmex André. *Bulletin of the British Museum (Natural History) (Entomology)* 32: 1-16. [10.iv.1975.]

Bolton, B. 1975c. The sexspinosa-group of the ant genus Polyrhachis F. Smith. Journal of Entomology (Series B) 44: 1-14. [27.v.1975.]

Bolton, B. 1976. The ant tribe Tetramoriini. Constituent genera, review of smaller genera and revision of Triglyphothrix Forel. *Bulletin of the British Museum* (*Natural History*) (*Entomology*),34: 281-379.

Bolton, B. 1977. The ant tribe Tetramoriini. The genus Tetramorium Mayr in the Oriental and Indo-Australian regions, and in Australia. *Bulletin of the British Museum* (*Natural History*) (*Entomology*) 36: 67-151.

Bolton, B. 1979. The ant tribe Tetramoriini. The genus *Tetramorium* Mayr in the Malagasy region and in the New World. *Bulletin of the British Museum* (*Natural History*) (*Entomology*), 38: 129-181.

Bolton, B. 1980. The ant tribe Tetramoriini. The genus *Tetramorium* Mayr in the Ethiopian zoogeographical region. *Bulletin of the British Museum* (*Natural History*) (*Entomology,*) 40: 193-384.

Bolton, B. 1981a. A revision of the ant genera *Meranoplus* F. Smith, Dicroaspis Emery and Calyptomyrmex Emery in the Ethiopian zoogeographical region. *Bulletin of the British Museum* (*Natural History*) (*Entomology*), 42: 43-81.

Bolton, B. 1981b. A revision of six minor genera of Myrmicinae in the Ethiopian zoogeographical region. *Bulletin of the British Museum* (*Natural History*) (*Entomology*), 43: 245-307.

Bolton, B. 1983. The Afrotropical dacetine ants. *Bulletin of the British Museum* (*Natural History*) (*Entomology*) 46: 267-416.

Bolton, B. 1984. Diagnosis and relationships of the Myrmicinae ant genus *Ishakidris* gen.n. *Systematic Entomology*, 9: 373-382.

Bolton, B. 1985. The ant genus Triglyphothrix Forel a synonym of Tetramorium Mayr. *Journal of Natural History* 19: 243-248. [23.v.1985.]

Bolton, B. 1986a. A taxonomic and biological review of the Tetramoriini ant genus *Rhoptromyrmex*. *Systematic Entomology*, 11: 1-17.

Bolton, B. 1986b. Apterous females and shift of dispersal strategy in the *Monomorium salomonis*-group. *Journal of Natural History*, 20: 267-272.

Bolton, B. 1987. A review of the *Solenopsis* genus-group and revision of Afrotropical *Monomorium* Mayr. *Bulletin of the British Museum* (*Natural History*) (*Entomology*), 54: 263-452.

Bolton, B. 1988a. A new socially parasitic Myrmica, with a reassessment of the genus. *Systematic Entomology* 13: 1-11. [(31).i.1988.]

Bolton, B. 1988b. A review of Paratopula Wheeler, a forgotten genus of myrmicine ants. *Entomologist's Monthly Magazine* 124: 125-143. [27.vii.1988.]

Bolton, B. 1988c. Secostruma, a new subterranean tetramoriine ant genus. *Systematic Entomology* 13: 263-270. [(31).vii.1988.]

Bolton, B. 1990a. Abdominal characters and status of the cerapachyine ants. *Journal of Natural History* 24: 53-68. [22.ii.1990.]

Bolton, B. 1990b. The higher classification of the ant sub-family Leptanillinae. *Systematic Entomology* 15: 267-282. [(31).vii.1990.]

Bolton, B. 1990c. Army ants reassessed: the phylogeny and classification of the doryline section. *Journal of Natural History* 24: 1339-1364. [19.x.1990.]

Bolton, B. 1991. New Myrmicinae ant genera from the Oriental region. *Systematic Entomology*,16: 1-13.

Bolton, B. 1992. A review of the ant genus *Recurvidris*, a new name for *Trigonogaster* Forel. *Psyche* 99: 35-48.

Bolton, B. 1994. Identification Guide to the Ant Genera of the World: 222 pp. Cambridge, Mass. [25.vii.1994; date author's copy received.]

Bolton, B. 1995a. A taxonomic and zoogeographical census of the extant ant taxa. *Journal of Natural History*,29: 1037-1056.

Bolton, B. 1995b. A New General Catalogue of the Ants of the World: 504 pp. Cambridge, Mass.

Bolton, B. 1998. Monophyly of the dacetonine tribe-group and its component tribes. *Bulletin of the Natural History Museum (Entomology Series)* 67: 65-78. [25.vi.1998.]

Bolton, B. 1999. Ant genera of the tribe Dacetonini. *Journal of Natural History*, 33: 1639-1689.

Bolton, B. 2000. The ant tribe Dacetini. *Memoirs of the American Entomological Institute* 65: 1028 pp. [28.xii.2000; date author's copy received.]

Bolton, B. 2003. Synopsis and classification of Formicidae. *Memoirs of the American Entomological Institute*, 71: 370pp.

Bolton, B. 2007a. Taxonomy of the Dolichoderinae ant genus *Technomyrmex* Mayr based on the worker caste. Contributions of the American Entomological Institute 35 (1): 1-150.

Bolton, B. 2007b. How to conduct large-scale taxonomic revisions in Formicidae. Memoirs of the American Entomological Institute 80: 51-71.

Bolton, B. 2014. "Formicidae". *AntCat*.Retrieved 15 September 2014.

Bolton, B., and Marsh, A.C. 1989. The Afrotropical thermophilic ant genus *Ocymyrmex*. *Journal of Natural History*,23: 1267-1308.

Bolton, B., 1995. A new general catalogue of the Ants of the World. Harvard University Press, Cambridge, Massachusetts, USA.

Bolton, B., Gotwald, W.H. and Leroux, J.-M. 1979. A new West African ant of the genus *Plectroctena* with ecological notes. *Annales de l'Université d'Abidjan. Série E (Ecologie,)*9 (1976): 371-381.

Bolton, B., Sosa-Calvo, J., Fernández, F. and Lattke, J.E. 2008. New synonyms in neotropical Myrmicinae ants. *Zootaxa*,1732: 61-64.

Boulton, A.M., Davies, K.F. and P.S. Ward, 2005. Species richness, abundance, and composition of ground-dwelling ant in Northern California grasslands: Role of plants, soil and grazing. *Environ. Entomol*, 34(1): 96-104.

Boursaux-Eude C, Gross R. New insights symbiotic associations between ants and bacteria. *Res Microbiol* 2000; 519:513-519.

Brown,SGA., Heddle,RJ., Wiese, MD and KE,Blackman,2005. "Efficacy of ant venom immunotherapy and whole body extracts". *The Journal of Allergy and Clinical Immunology* **116** (2): 464-465. doi:10.1016/j.jaci.2005.04.025. PMID 16083810.

Brown, Jr. W. L, 1958. A review of the Ants of New Zealand (Hymenoptera). *Acta Hymenopterologica*, 1 (1): 1-50.

Brown, W. L. and W. W. Kempf, 1969. A revision of the Neotropical Dacetine ant genus *Acanthognathus* (Hymenoptera: Formicidae). *Psyche*, 76(2): 87-109.

Bucher (1982). Chaco and Caatinga – South American Arid Savannas, Woodlands and Thickets. In Ecology of Tropical Savannas (B.J.Huntely and B.H.Walker, eds):48-79.

Bueno OC, Fowler HG. Exotic ants and native ant fauna of Brazilian hospitals. *In*: Williams DF, editor. Exotic ants: Biology, impact and control of introduced species. Boulder: Westview Press; 1994. p. 191-198.

BVIEER(2010).Current Ecological Status and Identification of Potential Ecologically Sensitive Areas in the Northern Western Ghats.Bharti Vidyapeeth Institute of Environment Education And Research, Pune.

Calcaterra, L.A, Vander Meer, R.K., Pitts, J.P. Livore, J.P. and N.D. Tsutsui, 2007. Survey of *Solenopsis* fire ants and their parasitoid flies (Diptera: Phoridae: *Pseudacteon*) in central Chile and Central Western Argentina. *Ann. Entomol. Soc. Am*, 100(4): 512-521.

Callcott, A.A., OI, D.H., Collins, H.L., Williams, D.F. and T.C. Lockely, 2000. Seasonal studies of an isolated red imported fire ant (Hymenoptera: Formicidae) population in Eastern Tennessee. *Environ. Entomol*, 29(4): 788-794.

Calvo. S.J. and T.R. Schultz. 2010. Three remarkable new fungus - growing ant species of the genus *Myrmicocrypta* (Hymenoptera: Formicidae), with a reassessment of the characters that define the genus and its position within the Attini. *Ann. Entomol. Soc. Am.* 103(2): 181-195.

Chapman, J. W. and Capco, S. R. 1951. Check List of the Ants (Hymenoptera: Formicidae) of Asia. *Monogr. Inst. Sci. Tech., Manila*, 1: 1-327, 1 map.

Chapman, RE., Bourke, Andrew FG,2001. "The influence of sociality on the conservation biology of social insects" (PDF). *Ecology Letters* **4** (6): 650–662. doi:10.1046/j.1461-0248.2001.00253.x.

Chatterjee, P.N. and M.P. Mishra, 1975. Natural insect enemy and plant lost complex of forest insects pests of Indian region. *Indian for Bull. No.265 (M.S.)* Entomology, Controller of Publications, Govt. of India, Delhi pp. 1-223.

Chavan, A. and S. S.Pawar, 2011. Distribution and diversity of ant species (Hymenoptera: Formicidae) in and around Amravati city Maharashtra, India. *World Journal of Zoology*, 6(4): 395-400.

Cherret, J.M, (1989). Leaf cutting ants. Biogeographical and ecological studies. In ecosystem of the world. Tropical Rain Forest Ecosystem (H.Leith and M.J.Werger, eds):473-488.

Clarke, P.S, 1986. "The natural history of sensitivity to jack jumper ants (Hymenoptera: Formicidae: *Myrmecia pilosula*) in Tasmania". *Medical Journal of Australia* **145** (11–12): 564–566. PMID 3796365.

Cunha, P.D. and V.M.G. Nair, 2013. Diversity and distribution of ant fauna in Hejamadi Kodi Sand spit, Udupi District, Karnataka, India. *Halters*, 4: 33-47.

Currie C.R., Scott J.A., Summerbell R.C., Malloch D. (1999) Fungus-growing ants use antibiotic producing bacteria to control garden parasites, *Nature* 398, 701–704.D. C. 280 pp.

Daniels, R, 2001 a. Biodiversity of the Western Ghats : An overview. In Research Priorities in Tropical Rain Fore STS in India. Vol. 2. Coimbatore : Wildlife Institute of India, SACON, State Forest College.

Daniels, RJ.R. 1991. Ants as biological indicators of environmental changes. *Blackbuck*, 7:51-56.

Das, Amalendu, 2003. Zoological Survey of India. A catalogue of new taxa described by the scientists of the zoological survey of India, during 1916-1991. The survey.

Dasmann, R. F., 1984. Environmental conservation New York.

DeFoliart, G.R. 1999. "Insects as food: Why the western attitude is important". *Annual Review of Entomology* **44**: 21–50. doi:10.1146/annurev.ento.44.1.21. PMID 9990715.

Dicke, E.,Byde, A., Cliff, D.,P. Layzell,2004. A. J. Ispeert, M. Murata and N. Wakamiya, ed. "An ant-inspired technique for storage area network design". *Proceedings of Biologically Inspired Approaches to Advanced Information Technology: First International Workshop, BioADIT 2004 LNCS 3141*: 364–379.

Diddee,J,2002.Geography of Maharashtra. Rawat Publications,Janury.

Donisthorpe, H.D., 1942. Ants from Colombo Museum Expedition of South India. *Ann. Mag. Mat. Hist. Landon* 9: 449-461

Dunn, R.R., Sanders, N.J., Fitzpatrick, M.C., Laurent, Ed., Lessard, J.P., Agosti, D., Andersen, A.N., Bruhl, C., Cerda, X., Ellison, A.M., Fisher, B.L., Gibb, H., Gotelli, N.J. Gove, A., Guenard, B., Janda, M., Kaspari, M., Longino, J.T., Majer, J., Mcglynn, T.P., Menke, S.B., Parr, C.L., Philpott, S.M., Pfeiffer, M., Retana, J., Suarez, A.V. and H.L.Vasconcelos, 2007. Global ant (Hymenoptera: Formicidae) biodiversity and biogeography a new database and its possibilities. *Myrmecological News*, 10: 77-83.

Edwards JP, Baker LF,1981.Distribution and importance of The Pharaoh's *Monomorium pharaonis* (L.) in national Health Service Hospitals in England.*J Hosp Infec*, 2:245-254.

Eguchi, K., Yoshimura, M. and Yamane, S. 2006. The Oriental species of the ant genus *Probolomyrmex*. *Zootaxa*, 1376: 1-35.

Emery, C, 1901.Notes sur les sous-familles des dorylines et ponerines. *Ann. Soc. Ent. Belg*, 45:32-54.

Emery, C, 1906. Notes sur *Prenolepis vividula* Nyl. Et sur la classification des especes du genre *Prenolepis*. *Ann. Soc. Ent. Belg*, 50: 130-134

Emery, C, 1910.Hymenoptera, Fam.Formicidae, sub-fam.Dorylinae. *Genara Insect.*, Brussels, *Fasc.102:1*-33, 1 pl.

Emery, C, 1911. In Wytsman, P. *Genera Insectorum*. Hymenoptera. Fam. Formicidae. sub-fam. Ponerinae. *Fasc*, 118:124pp.

Emery, C, 1912. Hymenoptera, Fam.Formicidae, sub-fam. Dolichoderinae. *Genara Insect.*, Brussels, *Fasc.137:1*-50, 2 pls.

Emery, C, 1921.Hymenoptera, Fam.Formicidae, sub-fam.Myrmicinae. *Genara Insect.*, Brussels, *Fasc*. 174 A: 1-94, pls.i-vii.

Erwin, T. L, 1989. Sorting tropical forest canopy samples (an experimental work for networking information). *Insect Collection News*, 2(1): 8.

Erwin, T.L. (1983). In: *Ecology and Management of Tropical Rain forest*.S.L. Sutton, T.C. Whitmore, and A.C. Chadwick (eds.). Oxford:Blackwell.

Fabricius, J.C, 1775. *Systema Entomologiae*, sistens insectorum classes, ordines, genera, species, adiectis synonymis, locis, descriptionibus, observationibus: 834pp.

Fabricius, J.C, 1787.Mantissa Insectorum, 1:348pp.

Fabricius, J.C, 1798. Supplementum Entomologiae Systematicae: 572pp.

Feldhaar, H.; Straka, J.; Krischke, M.; Berthold, K.; Stoll, S.; Mueller, M.J.; Gross, R. (2007). "Nutritional upgrading for omnivorous carpenter ants by the endosymbiont Blochmannia". *BMC Biology* **5**: 48. doi:10.1186/1741-7007-5-48. PMC 2206011.PMID 17971224.

Fischer, G., Hita Garcia, F. and Peters, M.K. 2012. Taxonomy of the ant genus Pheidole Westwood in the Afrotropical zoogeographic region: definition of species groups and systematic revision of the Pheidole pulchella group. *Zootaxa*, 3232: 1-43.

Fischer, R.C., Ölzant, S.M., Wanek, W and V. Mayer,2005. "The fate of *Corydalis cava*elaiosomes within an ant colony of *Myrmica rubra*: elaiosomes are preferentially fed to larvae". *Insectes sociaux* **52** (1): 55–62. doi:10.1007/s00040-004-0773-x.

Fisher, B.L. 1997. Biogeography and ecology of the ant fauna of Madagascar. *Journal of Natural History*, 31: 269-302.

Fittkau, E.J. and H. Klinge, 1973 On biomass and tropic structure of the Central Amazonian rain forest ecosystem. *Biotropica*, 5:2-14.

Folgarait, P.J, 1998. Ant biodiversity and its relationship to ecosystem functioning: a review. *Biodiversity and Conservation*, 7:1221-1244.

Forel, A, 1892. Nouvelles especes de formicides de Madagascar. *Ann.Soc. Belg*, 36:516-535.

Forel, A, 1894. Les formicides de I ' Empire des Indes et de Ceylan. Part 4. Adjonction aux genes *Camponotus* Mayr, et Polyrachis Shuck. *J. Bombay Nat. Hist.Soc*, 8:396-420.

Forel, A, 1895. Les formicides de I ' Empire des Indes et de Ceylan. Part 5. Adjonction aux genes Camponotinae. *J. Bombay Nat. Hist.Soc*, 9:433-472.

Forel, A, 1901. Les formicides de I ' Empire des Indes et de Ceylan. Part 8. Sous famille Dorylinae. *J. Bombay Nat. Hist.Soc*, 13:462-477.

Forel, A, 1902. Myrmicinae nouveaux de I' Indes et Ceylan. *Rev. Suiss. Zool*, 10:165-249.

Forel, A, 1903.Les Formicides De L' empire des Indes et de Ceylan. *J. Bombay. Nat. Hist. Soc*, 10:679-715.

Fowler HG, Bernardi OC, Sadatsube T, Montelli AC. Ants as potencial vectors of pathogens in Brazil hospitals in the State of São Paulo, Brazil. *Insec Scie Applic* 1993; 14:367-370.

Francisco, H. G., Eli, M. Sarnat and E. P. Economo,(2015). Revision of the ant genus Proceratium Roger (Hymenoptera, Proceratiinae) in Fiji. *Zookeys*,475:97-112.

Franks, NR., Hooper, J., Webb, C., and A. Dornhaus,2005. "Tomb evaders: house-hunting hygiene in ants". *Biology Letters* **1** (2): 190–192. doi:10.1098/rsbl.2005.0302.PMC 1626204. PMID 17148163.

Friedrich, Russell., Philpott, Stacy M,2009. "Nest-site Limitation and Nesting Resources of Ants (Hymenoptera: Formicidae) in Urban Green Spaces". *Environmental Entomology*, **38** (3): 600–607. doi:10.1603/022.038.0311.

Frouz, J, 2000. "The Effect of Nest Moisture on Daily Temperature Regime in the Nests of *Formica polyctena* Wood Ants.". *Insectes Sociaux* **47**: 229–235.doi:10.1007/PL00001708.

Gadagkar, R., Nair, P., Chandrashekara, K and D.M.Bhat, 1993. Ant species richness and diversity in some selected localities of Western Ghats, India. *Hexapoda*, 5(2):79-94.

Galle, L, 1972. Study of ant population in various grass land ecosystem. *Acta Biologica* 18: 159-164.

Gaonkar, H, 1996. The butterflies of Western Ghats India and Srilanka. A biodiversity assessment on the ecological history of Western Ghat Abst., pp.99.

Gatson, K.J and Spicer, J.J. 2004. Biodiversity : an introduction, 2nd edn. Blackwell Publishing, Oxford, UK.

Ghosh, S.N., Sheela, S. and Kundu, B.G. 2005. Ants of Rabindra Sarovar, Kolkata. Records of the Zoological Survey of India, Occasional Papers 234: 1-40.

Giladi,I, 2006. "Choosing benefits or partners: a review of the evidence for the evolution of myrmecochory". *Oikos* **112** (3): 481–492. doi:10.1111/j.0030-1299.2006.14258.x.

Gillott, C, 1995. *Entomology*. Springer. p. 325. ISBN 0-306-44967-6.

Girard, M, 1879. Traite elementaire *d 'entomology*, 2: 1028 pp.

Goss, S., Aron, S., Deneubourg, J.L and J.M. Pasteels, 1989. "Self-organized shortcuts in the Argentine ant". *Naturwissenschaften* **76** (12):579 581. Bibcode:1989NW.76.579G. doi:10.1007/BF00462870. doi:10.1007/BF01012125.

Groc, S., Delabie, JH.C. Cereghino, R., Orivel, J., Jaladeau, F., Grangier, J., Mariano, CS.F. and A. Dejean, 2007. Ant species diversity in the 'Grands Causses' (Aveyron, France): In search of sampling methods adapted to temperate climates. *C.R.Biologies*, 330: 913-922.

Guerin-Meneville F.E, 1844. Iconographie du regne animal de G. Cuvier, Vol.7, Insects. Bailliere Brothers, Paris, 576pp.

Haddad, Jr. V., Cardoso,JLC and RHP Moraes, 2005. "Description of an injury in a human caused by a false tocandira (*Dinoponera gigantea*, Perty, 1833) with a revision on folkloric, pharmacological and clinical aspects of the giant ants of the genera Paraponera and Dinoponera (sub-family Ponerinae)" (PDF). *Revista do Instituto de Medicina Tropical de São Paulo* **47** (4): 235–238. doi:10.1590/S0036-46652005000400012.

Haneda, N.F., Sajap, A.S. and M.Z.Hussin, 2005. A study of two ant (Hymenoptera: Formicidae) Sampling Methods in Tropical Rain Forest. *Journal of Applied Sciences*, 5(10): 1732-1734.

Hanzawa, F.M., Beattie, A.J and D.C, Culver, 1988. "Directed dispersal: demographic analysis of an ant-seed mutualism". *American Naturalist* **131** (1): 1–13. doi:10.1086/284769.

Hölldobler and Wilson (1990a), p. 573.

Hölldobler and Wilson (1990b), pp. 143–179.

Hölldobler and Wilson (1990c), pp. 351, 372.

Hölldobler and Wilson (1990d), pp. 619–629.

Hölldobler B., Wilson E.O. (1990) The ants. Springer,Heidelberg, 732 p.

Holldobler B., Wilson E.O. (2009). The superorganism, W.W.Norton and Company, Inc., New York, 522 p.

Holldobler, B and E. O.Wilson, 1990. The Ants. Harvard University Press. U.S.A.pp.1-732.

Holldobler, B. and Wilson, E.O. (1990) *The Ants.* Massachusetts:Harvard University Cambridge, Belknap Press.

Holldobler, B. and Wilson, E.O. (2009) *The Super-Organism: TheBeauty elegance, and Strangeness of Insect Societies.* New York–London: W.W. Norton and Company.

Holway, D.A., Lach, L., Suarez, A.V., Tsutsui, N.D. and Case, T.J.(2002) The causes and consequences of invasions. *AnnualReview of Ecology and Systematics*, **33**: 181-233.

Hughes, J. 2006. A review of wood ants (Hymenoptera: Formicidae) in Scotland. *Scottish Natural Heritage* Commissioned Report No. 178.

Hung, A.C.F. 1970. A revision of ants of the subgenus *Polyrhachis* Fr. Smith. *Oriental Insects*, 4: 1-36.

ITO, E., Yamane, S., Egychi, K., Noerdjito, W.A., Kahono, S., Tsuji, K., Ohkawara, K., Yamauchi, K., Nishida, T and K, Nakamura, 2001. Ant species diversity in the Bogor botanic garden, West Java, Indonesia, with descriptions of two new species of the genus *Leptanilla* (Hymenoptera, Formicidae). *Tropics,* 10(3): 379 - 404.

Jackson,D.E.and F.L.Ratnieks,August 2006. "Communication in ants". *Curr. Biol.* **16**(15): R570–R574. doi:10.1016/j.cub.2006.07.015. PMID 16890508.

Jadav Divya K. and Sathe T. V. 2014. Altitudinal diversity of forensic Blow flies (Diptera: Calliphoridae) of Western Ghats (Maharashtra). *J. Forensic Res.*, 5, 251.

Jadav Divya K. and Sathe T. V. 2015. Diversity of forensic Blow flies (Diptera: Calliphoridae) from Western Ghats, Maharashtra, India. *Indian journal of applied research,* 5(9), 55-60.

Jadav Divya K. and Sathe T. V. 2015. Diversity, occurrence and development of forensic insects on Dog (*Canis domesticus* L.) carcass from Kolhapur, India. *International Journal of Pharma & Biological sciences,* 6(4)(B), 498-506.

Jadhav A. D., Desai A. S. and T. V. Sathe. 2014. Distribution and economic status of Uzi Fly *Exorista Bombycis* Louis, a parasitoid of mulberry silk worm *Bombyx Mori* L. in Maharashtra. *Global journal for research analysis*, 3 (7), 3- 5.

Jadhav, B.V. and T.V. Sathe, 2006a. Biodiversity of aphid (order- Hemiptera) from Satara district of Western Ghats. *Indian J. Environ. and Ecoplan.*, 12(1), 237-240.

Jadhav, B.V. and T.V. Sathe, 2006b. Biodiversity of aphid (Order - Hemiptera) from Poona district of Western Ghats. *J. Adv. Zool.*, 27(1), 43-45.

Jahn,G.C and J. W. Beardsley,1996. "Effects of *Pheidole megacephala* (Hymenoptera: Formicidae) on survival and dispersal of *Dysmicoccus neobrevipes* (Homoptera: Pseudococcidae)". *Journal of Economic Entomology*, **89**: 1124–1129.

Jaitrong, W and S, Yamane, 2012. Review of the Southeast Asian species of the *Aenictus javanus* and *Aenictus philippinensis* species groups (Hymenopter, Formicidae and Aenictinae). *Zookeys*, 192:49-78. Doi: 10. 3897/*Zookeys*, 193.2768.

Jaitrong, W., Yamane, S and D, Wiwatwitaya, 2010. The army ant *Aenictus wroughtonii* (Hymenoptera: Formicidae: Aenictinae) and related species in the Oriental Region, with description of two new species. *Japanese Journal of Systematic Entomology*, 16:33-46.

Jerdon, T.C, 1851. A catalogue of the species of Ants found in Southern India. *Madras J. Lit. Sci*, 17:103-127.

Joseph, T.M., 2004. Biodiversity conservation in the Western Ghat. *Biodiversity and Environment*, 17, 221-227.

Kale, V.S.2010. The Western Ghat : The Great Escarpment of India. In : *Geomorphological Landscapes of the World.* Pioter Migon (Ed.), Springer, New York, USA, PP. 257-264.

Kamatar, V.K, 1983. Ants of Raichur District (Karnataka State) with observation on the Biology and Behavior of fire ant *Solenopsis geminate* Fabricius. M.Sc. Thesis, USA, Bangalore.

Kannowski, P. B, 1957.Notes on the ant *Leptothorax provancheri* Emery. *Psyche*, 64(1): 1-5.

Karmaly, K.A. 2004. A new species and a key to species of *Polyrhachis* Smith from India (pp. 539-551). In, Rajmohana, K., Sudheer, K., Girish Kumar, P. and Santhosh, S. (eds). Perspectives on biosystematics and biodiversity. Prof. T.C. Narendran commemorative volume. Kerala: *Systematic Entomology Research Scholars Association*: 666 pp.

Karmaly, K.A. and Narendran, T.C. 2006. Indian ants: genus *Camponotus*: 165 pp. Teresian Carmel Publications, Kerala.

Kataria. R. and D. Kumar, 2013. On the apid - ant association and its relationship with various host plants in the agroecosystems of Vadodara, Gujarat, India. *Halters*, 4: 25-32.

Kavane, R.P. and T.V. Sathe, 2011. Wild silk technology. *Daya publishing house, New Delhi*, pp.1-234, ISBN-948-81-7035-712-4.

Keller, L,1998. "Queen lifespan and colony characteristics in ants and termites". *Insectes Sociaux* **45** (3): 235–246. doi:10.1007/s000400050084.

Kharbani, H and S.R. Hajong, 2013. Seasonal patterns in ant (Hymenoptera: Formicidae) activity in a forest habitat of the West Khasi Hills, Meghalaya, India. *Asian Myrmecology,* 5: 103-112.

Khot, K., Quadros, G and V. Somani, 2013. Ant diversity in an urban garden at Mumbai, Maharashtra. *National Conference on Biodiversity: Status and Challenges in Conservation* - 'FAVEO'.

Kistner, D. H. (1982). *Social Insects III.* H. R. Hermann (ed.). NewYork: Academic Press.

Krivokhatskii, V.A, 1997. Some little known and a new species of Ant-Lions (Neuroptera, Myrmeleontidae) from Indo-China. *Entomological Review,* 77(7): 807-814.

Kumar Sunil M.K.T., Nair P., Varghese, T and R. Gadagkar, 1997. Ant species richness at selected localities of Bangalore. *Insect Environment,* 3:3-5.

Kumar, D and A. Mishra, 2008. Ant community variation in Urban and agricultural ecosystems in Vadodara District (Gujarat State), Western India. *Asian Myrmecology,* 2: 85-93.

Kunte, K.J. 1997, Seasonal patterns in butterfly abundance and species diversity in four tropical habitats in northern Western Ghats. *Journal of Biosciences,* 22(5): 593-603.

Kurane, S., Bhoje, P.M and T.V. Sathe, 2015. Diversity and Economic Importance of Ants (Hymenoptera: Formicidae) From Kolhapur City, Maharashtra State, India. *International Journal of Development Research,* 5(3):3662-3664.

Lach, L., Parr, C.L. and Abbott, K.L. (2010). *Ant Ecology.* Oxford:University Press.

Lachaud, J.P. and G.P. Lachaud, 2012. Diversity of Species and behavior of Hymenoptera Parasitoids of Ants: A review. *Psyche.* pp.24.

Lange, D., Fernandez, W. D., Raizer, J and O, Faccenda, 2008. Predacious activity of ants (Hymenoptera: Formicidae) in conventional and in no-till agriculture systems. *Brazilian Archives of Biology and Technology,* 51(6):1199-1207. ISSN: 1516-8913.

Latreille, P.A, 1802. Histoire Naturelle des Fourmi, et recuteil de memoires et d' observations sur les abeilles, les araignees, les faucheurs, et autres insects, Paris.445 pp. *Formica longipes* Latreille, 1802:233. Original description.

Linnaeus, C, 1758. Systema Naturae per regna tria naturae, Secundum classes, ordines, genera, species, cum, characteribus, differentilis, synonymis, locis. Editio, 10, 1:823 pp.

Linnaeus, C, 1764.Museum Sae Rae Mtis Ludovicae Ulricae Reginae Svecorum, Gothorum, Vandalorumque. In quo animalia rariora, exotica, imprimis. Insecta and Conchilia describuntur and determinantur. *Prodromi Instar*: 719pp.

Lobry de Bryuyn L.A. (1999). Ants as bioindicators of soil function in rural environments, *Agr. Ecosyst. Environ.* 74,425–441.

Longino, J.T, (2006). A taxonomic review of the genus *Myrmelachista* (Hymenoptera: Formicidae) in Costa Rica. *Zootaxa*, 1141: 1-54. ISSN: 1175-5326, ISSN: 1175-5334.

Longino, J.T., Codington, J and R.K.Colwell, 2002. The ant fauna of a tropical rain forest: estimating species richness three different ways. *Ecology*, 83(3): 689-702.

Majer JD. 1990. The abundance and diversity of arboreal ants in Northern Australia. *Biotropica*, 22: 191 – 199.

Mayr, G. 1861. Die Europdischen Formiciden. (Ameisen): 80 pp. Wien.

McGain, F., and K.D.Winkel,2002. "Ant sting mortality in Australia". *Toxicon* **40** (8):1095–1100. doi:10.1016/S0041-0101(02)00097-1. PMID 12165310.

MoEF (2014). India's Fifth National Report to the Conservation on Biological Diversity. Ministry of Environment and Forest, Government of India.

Mortazavi,Z.S., Sadeghi,H., Aktac,N., Depa,L and L, Fekrat, 2015. Ants (Hymenoptera: Formicidae) and their aphid partner (Homoptera: Aphididae) in Mashhad region. Razavi Khorasan province with new records of aphids and ant species for Fauna of Iran. *Halters*, 6:4-12. ISSN: 0973-1555, ISSN: 2348-7372.

Mueller, U. G., Schultz, T. R., Currie, C. R., Adams, R. M. M. andMalloch, D. (2001) The origin of the attine ant-fungussymbiosis. *Quarterly Review of Biology*, **76**: 169-197.

Nagariya, S.A. and S.S. Pawar, 2012. Distribution of (Hymenoptera: Formicidae) ants diversity in Pohara forest area of Amravati Region, Maharashtra State, India. *International Journal of Science and Research*, 3(7): 1310-1312.

Nair, K.S., Mathew, G. and M. Shivarajan, 1973. Occurence of bogworm *Plagiosphleps* Hampsen (lepidoptera : Psychidea) as a pest of the free *Albizia falcataria* in Kerala, India. *Entomon*, 6, 179-180.

Narendra A (2003) Responses of the Asian Weaver Ant, *Oecophylla smaragdina* towards high quality and quantity food substances. *Insect Environment*, 9: 89-90.

Viswanathan G, Narendra A (2000) Food preference in different species of ants. *Insect Environment*, 6:34-35.

Neves, F.S., Braga, R.F., Espirito-Santo, M.M., Delabie, J.C., Fernandes, G.W and G.A, Sanchez-Azofeifa, 2010. Diversity of arboreal ants in a Brazilian tropical dry forest: Effects of seasonality and successional stages. *Sociobiology*, 56(1):1-18.

Nylander, W, 1846. Adnotationes in monographiam formicarum borealium Europae. *Acta Soc. Sci. Fenn*, 2: 875-944.

Obin, M.S and R.K.Vander Meer,1985. "Gaster flagging by fire ants (*Solenopsis spp.*): Functional significance of venom dispersal behavior". *Journal of Chemical Ecology*, 11(12): 1757–1768.

Ogata, K and H. Okido, 2007. Revision of the ant genus *Perissomyrmex,* with notes on the phylogeny of the tribe Myrmecinini. *Memoirs of the American Entomological Institute,* 80: 352-369.

Olson, J.S., Watts, J.A., Allison, L.J. 1983 Carbon in live vegetation of major world ecosystem. Oak Ridge National Laoratory : WC MC 1992.

Oster, G. F and E.O.Wilson, 1978. *Caste and ecology in the social insects.* Princeton University Press, Princeton. pp. 21–22. ISBN 0-691-02361-1.

Pandharbale, A.R. and T.V. Sathe, 2001. On a new species of the Genus *Syntomis* (Syntomidae : Lepidoptera) from the environment of Western Ghats (Satara District). *Indian J. Environ. and Ecoplan.,* 5(2), 601-602.

Patil V. S. and Sathe T. V. 2003. Predators & pest management. Daya publishing house, New Delhi. Pp. 1- 216.

Patkar N.B., B.V. Sonune and R. J. Chavan 2013, Studies on efficiency of ant collection methods in and around Great Indian Bustard Wildlife Sanctuary, Maharashtra state, India. *Indian Streams Research Journal, 3* (06). ISSN: 2230-7850.

Patkar, N.B., B.V.Sonune and R.J.Chavan and A.M.Gaikwad, 2014. Taxonomic study of ants (Myrmiconae: Formicidae).

Patkar, N.B., B.V.Sonune and R.J.Chavan, 2013. Studies on Ants (Hymenoptera: Formicidae) in and around Solapur city (M.S.) India. *Indian Streams Research Journal,* 3(7).ISSN:2230-7850.

Peck, L.S., Mcquaid, B. and Campbell, C.L. (1998). Using ant species (Hymenoptera: Formicidae) as a biological indicator of agro ecosystem condition. *Environmental Entomology,* **27**(5): 1102-1110.

Philpott. S.M. Perfecto, I and J. Vandermeer, 2006 Effects of management intensity and season on arboreal ant diversity and abundance in coffee agro ecosystems. *Biodiversity and Conservation,* 15: 139-155.

Pitts, J and TL. Pitts - Singer, 2001. A new host record for *Pseudacteon crawfordi* (Diptera: Phoridae). *Florida Entomologist,* 84(2).

Rabeling, C and M. Bacci, 2010. A new Workerless inquiline in the Lower Attini (Hymenoptera: Formicidae), with a discussion of social parasitism in fungus growing ants. *Systematic Entomology,* 35: 379-392.

Radchenko, A, 2003. Perissomyrmex nepalensis sp.nov.- new evidence of Old World origins for the genus (Hymenoptera, Formicidae). *Entomologica Basiliensia,* 25:13-22. ISSN: 0253-24834.

Radchenko, A, 2004. A review of the ant genera *Leptothorax* Mayr and *Temnothorax* Mayr (Hymenoptera, Formicidae) of the Eastern Palaearctic. *Acta Zoologica Academiae Scientiarum Hungaricae,* 50(2): 109-137.

Radchenko, A., Elmes, G.M. and B.T.Viet, 2006. Ants of the genus *Myrmica* (Hymenoptera: Formicidae) from Vietnam, with a description of a new species. *Myrmecologische Nachrichten,* 8: 35-44.

Rajagopal,T., Sevarkodiyone,SP and A. Manimozhi,2005.Ant diversity in some selected localities of Sattur taluk, Virudunagar District, Tamil Nadu. *Zoos Print Journal*, 20:1887-1888.

Ramesh, T., Hussain, K.J., Satpathy, K.K., Selvanayagam, M and M.V.R. Prasad, 2010. Diversity, distribution and species composition of ants fauna at Department of Atomic Energy (DAE) campus Kalpakkam, South India. *World Journal of Zoology*, *5*(1): 56-65.

Rastogi, N., Nair, P., Kolatkar, M., William, H and R.Gadagkar, 1997. Ant fauna of the Indian Institute of Science Campus –Survey and some preliminary observations. *J. Indian Inst.Sci.*77:133-140.

Reddy, D.N.R., Viswanath.B.N and V.V.Belavadi, 1981. A preliminary study of ant fauna of Bisthenhatti, Dharwad.In: *Progress in Soil Biology and Ecology in India. (Ed.)* G.K.Veeresh U.S.A.technical Series, 37:200-205.

Resende, J.J., Peixoto, P., Silva E.N., Delabie, JHC and GMM, Santosh, 2013. Arboreal ant assemblages respond differently to food source and vegetation physiognomies: a study in the Brazilian Atlantic Rain Forest. *Sociobiology*, 60(2): 174-182.

Robso N.S.K. and R.J. Kohout, 2005. "Evolution of nest-weaving behaviour in arboreal nesting ants of the genus *Polyrhachis* Fr. Smith (Hymenoptera: Formicidae)". *Australian Journal of Entomology*, **44** (2): 164–169. doi:10.1111/j.1440-6055.2005.00462.x.

Roders,W.A. and Panwar,H.S.1988. Planning a Wildlife Protected Areas Network in India,Volumes 1 and 2.Department of Environment and Forests and Wildlife Institute of India,Dehradun.

Roders,W.A., Panwar,H.S., and Mathur,V.B.2002. Wildlife Protected Areas in India: a Review(Executive Summary). Wildlife Institute of India,Dehradun.

Roy, P.S., Kushwaha, S.P.S., Murthy, M.S.R., Roy, A., Kushwaha, D., Reddy, C.S., Behera, M.D., Mathur, V.B., Padalia, H., Saran, S., Singh, S., Jha, C.S., and Porwal, M.C, 2012.Biodiversity Characterisation at Landscape Level:National Assessment.Indian Institute of Remote Sensing, Dehradun, India.140 pp., ISBN 81-901418-8-0.

Russell, J.A., Moreau, C.S.,Goldman-Huertas, B., Fujiwara, M., Lohman, D.J., and N.E. Pierce, 2000. "Bacterial gut symbionts are tightly linked with the evolution of herbivory in ants". *PNAS* **106**:21236–21241. doi:10.1073/pnas.0907926106.

Sabu,T.K., P.J.Vineesh and K.V.Vinod,2008. Diversity of forest litter-inhabiting ants along elevations in the Wayanad region of the Western Ghats. *Journal of Insect Science*,8:1-14.

Sapolsky, Robert M. 2001. *A Primate's Memoir: A Neuroscientist's Unconventional Life Among the Baboons*. Simon and Schuster. p. 156. ISBN 0-7432-0241-4.

Sathe T. V. 2007. Biodiversity of Wild Silkmoths From Western Maharashtra, India. *Bull. Ind. Acad. Seri.*, 2(1), 21-24.

Sathe T. V. 2008. Mass production Technique for *Campoletis chlorideae* (Uchida). *Biotechnological Approaches in Entomology*, 3, 64-74.

Sathe T. V. 2012. Biodiversity of Ichneumonid flies (Hymenoptera: Ichneumonidae) from Western Ghats, Maharashtra. *The scientific Temper*, 3 (1 & 2), 47-50.

Sathe T. V. 2012. Pests of Ornamental Plants. Daya publishing house, New Delhi. pp. 1-199.

Sathe T. V. 2013. Dragon flies production technology. Daya publishing house, New Delhi. Pp.1-102.

Sathe T. V. 2014. Ecology, epidemiology and control of Sand flies from Kolhapur region, India.

Sathe T.V. 2009. A textbook of forest Entomology. Daya publishing house, New Delhi. Pp. 1-234.

Sathe T.V. 2010. Biocontrol approaches in mulberry pest management. *Application of Biotechnology in Sericulture*, 21, 229-241.

Sathe T.V. 2010. Biodiversity of Damselflies (Odonata) from Koyna Dam and around area. *Flora & Fauna*, 16, 68-72.

Sathe T.V.2011. Ecology of Mosquitoes from Kolhapur district, India. *International Journal of Pharma & Biosciences*, 2 (4) (B), 103-111.

Sathe, T. V.2012. Biodiversity of tachinid flies. (Diptera: Tachinidae) from western Maharashtra. *International Journal of Plant Protection*, 5, 368-370.

Sathe, T.V. 1992. Natural enemies of some insect pests of economic importance. *Oikoassay*, 9, 15-17.

Sathe, T.V. 2003. Biodiversity of braconid pest biocontrol agents from Western Maharashtra. *Bull. Bio. Sci.*, 1, 73-75.

Sathe, T.V. and M.K. Mulla, 1995. Insect pests of Mulbery from Amboli, *Bikoassay*, 12 (1-2), 9.

Sathe, T.V. 1992, Fauna of aphids on plourts of economic importance found in Western Maharashtra, India. *J.Curr.Biosci.* 9(1), 27-31.

Sathe, T.V. and A.R. Pandharbale, 1999. Hawk moth (Sphing idae : Lepidoptera) from Western Ghats of Satara district, India *Bull. Biol. Sci.* : 1, 81-88.

Sathe, T.V. and K.P. Shinde, 2006. Diversity of butterflies from Western Ghats (Kolhapur district) *J.Nat.Con. 81(1)*, 181-184.

Sathe, T.V., Inamdar S.A. and M.V. Santhakumar, 1986-87. Fauna of butterflies from Western Maharashtra and Western Ghats (Part of Maharashtra only) *Indian J. Shivaji Uni. (Science)*, 23, 391-398.

Satish, P.M. 1996. Moths and butterflies of Bhadra Projects. M.Sc. Thesis submitted to Kumempy University, Shimoga, pp.80.

Saunders, W.W, 1842. Descriptions of two hymenopterous insects from northern India. *Trans. Ent.Soc.Lon*, 3:57-58(1841).

Savitha, S., Barve, N. and P. Davidar, 2008. Response of ants to disturbance gradients in and around Bangalore, India. *Tropical Ecology,* 49(2): 235-243.

Schaal, Stephan (27 January 2006). "Messel". *Encyclopedia of Life Sciences*.doi:10.1038/npg.els.0004143. ISBN 0-470-01617-5.

Schuckard, W.E, 1840. Monograph of the Dorylidae, a family of the Hymenoptera. *Ann. Nat. Hist,* 5: 315-328.

Schultz, T. R. and McGlynn, T. P. (2000) in *Ants: standard methods formeasuring and monitoring biodiversity.* D. Agosti, J. Majer, E.L.Alonso and T.R. Schultz (eds). Washington, DC: SmithsonianInstitution Press.

Schultz, T.R, 1999. "Ants, plants and antibiotics". *Nature* **398** (6730): 747–748. Bibcode:1999Natur.398.747S. doi:10.1038/19619.

Schultz, T.R, 2000). "In search of ant ancestors". *Proceedings of the National Academy of Sciences* **97** (26): 14028–14029. Bibcode:2000PNAS.9714028S.doi:10.1073/pnas.011513798. PMC 34089. PMID 11106367.

Selvarani, S., Amutha, C and P.V. Moorthi, 2014. Seasonal assemblage of leaf litter ants in Megamalai, Western Ghats, India. *International Journal of Environmental Biology,* 4(2): 196-200.

Shattuck,S.O,1999. *Australian ants: their biology and identification*. Collingwood, Vic: CSIRO. p. 149. ISBN 0-643-06659-4.

Sheela S. and T. C. Narendran. 1998. On five new species of Tetramorium (Hymenoptera: Formicidae) from India ENTOMON 23(1): 37-44.

Sheela, S and S. N. Ghosh. 2009. *Lophomyrmex changlangensis,* a new species of ant (Hymenoptera: Formicidae) from India. *Biosystematica,* 2(2): 17-20.

Sheela, S and T.C. Narendran. 1997. A new genus and a new species of myrmecinae (Hymenoptera: Formicidae from India. *J.Ecobio.*9:87-91.

Sheela, S, (2008). Handbook on Hymenoptera: Formicidae, Zoological Survey of India. 55pp.

Sheela, S. 2008. First record of the rare ant *Paratopula ceylonica* Wheeler (Hymenoptera: Formicidae) from Uttar Pradesh, India with a note to the genus. *Journal of Experimental Zoology,* 11(2):423-425.

Smith, F, 1853. Monograph of the genus *Cryptocerus,* belonging to the group Cryptoceridae-family Myrmicidae- division Hymenoptera Heterogyna. *Trans. Ent. Soc. Lon* (2), 2: 213-228.

Smith, F, 1858. Catalogue of Hymenopterous insects in the collection of the British Museum, 6: *Formicidae*: 216 pp.London.

Smith, F, 1877. Descriptions of new species of the genera *Pseudomyrma* and *Tetraponera* belonging to the family Myrmicidae. *Trans.Ent. Soc.* Lond, (4) 10:57-72.

Sonune B. V., R. J. Chavan (2013) Taxonomy and diversity of Ant genus *Tetraponera* (Pseudomyrmicinae: Formicidae) from forest area of Aurangabad district, Maharashtra India. *Jr. Deccan Current Science,* Vo 9 (1), ISSN: 0975-3044, pp. 160-165.

Sonune B.V and Chavan R. J. 2012 Taxonomic study of ants (Hymenoptera: Formicidae) around Gautala Autramghat sanctuary Aurangabad (M.S.) *Jr. Life Science Bulletin*, 9(1), ISSN: 0973-5453.

Sonune B.V. and R. J.Chavan (2011), Studies on diversity and species richness of ants (Hymenoptera: Formicidae) around Autramghat in Aurangabad district of Maharashtra, India. *Jr. Deccan Current Science*, 6. ISSN: 0975-3044, pp. 344-347.

Sonune, B.V and R, J, Chavan, 2011. Studies on diversity and species richness of ants (Hymenoptera: Formicidae) around Autramghat in Aurangabad district of Maharashtra, India. *Jr. Deccan Current Science*, 6. ISSN: 0975-3044, 344- 347.

Sonune, B.V and R. J. Chavan, 2012. Taxonomic study of ants (Hymenoptera: Formicidae) around Gautala Autramghat Sanctuary Aurangabad (M.S.) *Jr.Life Science Bulletin*, 9(1). ISSN: 0973-5453.

Sonune, B.V and R.J. Chavan, 2013. Taxonomy and diversity of Ant genus *Tetraponera* (Pseudomyrmicinae: Formicidae) from forest area of Aurangabad district, Maharashtra, India. *Jr. Deccan Current Science*, 9(1).ISSN:0975-3044, 160-165.

Stabbing, E.P,1940. Indian forest insects of economic importance, Coleoptera, J.K.J. Brothers, Bhopal, pp.1-648.

Stafford,C.T, 1996. "Hypersensitivity to fire ant venom". *Annals of allergy, asthma, and immunology* 77 (2): 87–99. doi:10.1016/S1081-1206(10)63493-X.

Stoll S., Gadau J., Gross R., Feldhaar H. (2007). Bacterial micro biota associated with ants of the genus *Tetraponera, Biol.J. Linn. Soc.* 90, 399–412.

Styrsky,J.D and Eubanks, M.D. (January 2007). "Ecological consequences of interactions between ants and honeydew-producing insects". *Proc. Biol. Sci.* **274** (1607):151164.doi:10.1098/rspb.2006.3701. PMC 1685857. PMID 17148245.

Tak, N and N.S, Rathore, 1996. Ant (Formicidae) Fauna of the Thar Desert In: *Faunal Diversity in the Thar Desert: Gaps in Research.*A.K.Ghosh, Q.H.Baqri and I.Prakash (Eds.).271-276.

Tak, N, 1995. Studies on Ants (Formicidae) of Rajasthan-I. Jodhpur.*Hexapoda*, 7:17-28.

Tinaut, A, Ruano, F and D.Martinez, 2005. Biology, distribution and taxonomic status of the parasitic ants of the Iberian Peninsula (Hymenoptera: Formicidae, Myrmicinae. *Sociobiology*, 46(2): 1-40.

Tiwari, R. N., Kundu, B. G., Sheela, S., Chowdhury S. Roy and S. N. Ghosh.2004. Insecta: Hymenoptera: Formicidae, Fauna of Manipur. Fauna of India Series: 605-626.

Tiwari, R.N, 1994. Two new species of a little known Genus *Myrmecina curtis* (Insecta: Hymenoptera: Formicidae) from Kerala, India.*Rec.Zool.Surv, India*, 2-4:151-58.

Tiwari, R.N. 1994a. Two new species of a little known genus *Myrmecina curtis* from Kerala, India. Record of the Zoological Survey of India 94 (2-4): 151-158. [1994.]

Tiwari, R.N, 1996. Taxonomic studies on Ants of Southern India (Insecta: Hymenoptera: Formicidae). *Memoirs*, 18:1-96

Tiwari, R.N. and Jonathan, J.K. 1986a. A new species of *Liomyrmex* Mayr from Andaman Islands. Record of the Zoological Survey of India,83: 87-90. [(31). vii.1986.]

Tiwari, R.N. and Jonathan, J.K. 1986b. A new species of *Metapone* Forel from Nicobar Islands. Record of the Zoological Survey of India,83: 149-153. [(31).vii.1986.]

Tiwari, R.N. 1999. Taxonomic studies on ants of southern India. Memoirs of the Zoological Survey of India 18: 1-96. [(31).vii.1999.]

Tschinkel, W.R. 1987. Seasonal life history and nest architecture of a winter-active ant, *Prenolepis imparis. Insects Sociaux,* 34(3): 143-164.

UNEP RRC.AP (2001). India,State of the Environment,Teri, New Delhi, India.

Varghese, T. (2003). Ants of the Indian Institute of Science Campus. Technical report No. 98, Centre for Ecological Sciences, Indian Institute of Science, Bangalore.

Varghese, T. (2004a). Record of *Strumigenys emmae* (Emery) (Formicidae: Myrmicinae) from Bangalore, Karnataka and a key to Indian species. *J.Bombay nat. hist.Soc,* 101, 170.

Varghese, T. (2004b). Taxonomic studies on ant genera of the Indian Institute of Science Campus with notes on their nesting habits. *Perspectives on Biosystematics and biodiversity T.C.N.Com.,* 485-502.

Varghese, T, (2006). A new species of the ant genus *Dilobocondyla* (Hymenoptera: Formicidae) from India, with notes on its nesting behavior. *Oriental Insects,* 40: 23-32.

Varghese, T. (2006). Description of a new species of the Ponerinae ant genus, *Emeryopone* (Hymenoptera: Formicidae) from Karnataka, India. *Biospectra,* 1: 89-92.

Varghese, T (2009). A Review of Extant Sub-families, Tribes and Ant Genera in India. *Biosystematica,* 3 (2): 81-89.

Varghese, T. An overview of the ant species diversity in India- with a special emphasis on Western Ghats (*In press*).

Varghese, T., Milind, M. and Gadagkar, R. (2004). Checklist of ants in the Insect Museum. Technical report No. 104. Centre for Ecological Sciences, Indian Institute of Science, Bangalore.

Vishnudas C.K. (2008) Woodpeckers and ants in India's shade coffee, *Indian Birds* 4, 9 - 11.

Viswanathan G, Narendra A (1999) A study of the behaviour of the ant *Myrmicaria brunnea* Saunders towards pheromones. *Insect Environment,* 5: 23-25.

Wang D., McSweeney K., Lowery B., Norman J.M. (1995) Nest structure of ant *Lasius neoniger* Emery and its implication to soil modification, *Geoderma* 66, 259–272.

Ward, P.S, 2001. Taxonomy, phylogeny and biogeography of the ant genus Tetraponera (Hymenoptera: Formicidae) in the Oriental and Australian regions. *Invertebrate Taxonomy,* 15:589-665.

Watt, A.D., Stork, N.E and B. Bolton, 2002. The diversity and abundance of ants in relation to forest disturbance and plantation establishment in Southern Cameron. *Journal of Applied Ecology*, 39: 18-30.

Wheeler, W. M, 1913. Ants, their structure, development and behavior. Columbia University Biological Series 9.

Wheeler, W.M, 1917. A list of Indian ants. Proceeding of the Indian academy of Science, 26:460-466.

Wilson, E.O, 1953. The origin and evolution of Polymorphism in ants. *Quarterly Review of Biology*, 28(2), 136-56.

Wilson E.O. 1987. The arboreal ant fauna of Peruvian Amazon forests: a first assessment. *Biotropica*, 19: 245-251.

Wilson, E.O. and Hölldobler, B,2005. "The rise of the ants: A phylogenetic and ecological explanation". *Proceedings of the National Academy of Sciences* **102** (21):74117414.Bibcode:2005PNAS.102.7411W. doi:10.1073/pnas.0502264102. PMC 1140440.PMID 15899976.

Zacharias, M. and Rajan, P.D. 2004a. *Discothyrea sringerensis*, a new ant species from India. *Zootaxa*, 484: 1-4.

Zacharias, M. and Rajan, P.D. 2004b. *Vombisidris humbolticola*: a new arboreal ant species from an Indian ant plant. *Current Science*, 87 (10): 1337-1338.

Zoological Survey of India 2012. Fauna of Maharashtra. State Fauna Series, 20(Part-2): 567-586.

Zungoli, P.A. and E. P. Benson, 2008. Seasonal occurrence of swarming activity and worker abundance of *Pachycondyla chinensis* (Hymenoptera: Formicidae) proceeding of the Sixth International Conference on Urban Pests.

Index

P

S

T

V

Z

www.ingramcontent.com/pod-product-compliance
Ingram Content Group UK Ltd.
Pitfield, Milton Keynes, MK11 3LW, UK
UKHW021952270726
14060UKWH00002B/473

9 789389 605136